The TWI Facilitator's Guide

How to Use
the TWI
Programs
Successfully

The TWI Facilitator's Guide

How to Use the TWI Programs Successfully

Donald A. Dinero

CRC Press
Taylor & Francis Group
Boca Raton London New York

CRC Press is an imprint of the
Taylor & Francis Group, an **informa** business

A PRODUCTIVITY PRESS BOOK

CRC Press
Taylor & Francis Group
6000 Broken Sound Parkway NW, Suite 300
Boca Raton, FL 33487-2742

© 2017 by Donald A. Dinero
CRC Press is an imprint of Taylor & Francis Group, an Informa business

No claim to original U.S. Government works

Printed on acid-free paper
Version Date: 20160419

International Standard Book Number-13: 978-1-4987-5484-2 (Paperback)

This book contains information obtained from authentic and highly regarded sources. Reasonable efforts have been made to publish reliable data and information, but the author and publisher cannot assume responsibility for the validity of all materials or the consequences of their use. The authors and publishers have attempted to trace the copyright holders of all material reproduced in this publication and apologize to copyright holders if permission to publish in this form has not been obtained. If any copyright material has not been acknowledged please write and let us know so we may rectify in any future reprint.

Except as permitted under U.S. Copyright Law, no part of this book may be reprinted, reproduced, transmitted, or utilized in any form by any electronic, mechanical, or other means, now known or hereafter invented, including photocopying, microfilming, and recording, or in any information storage or retrieval system, without written permission from the publishers.

For permission to photocopy or use material electronically from this work, please access www.copyright.com (http://www.copyright.com/) or contact the Copyright Clearance Center, Inc. (CCC), 222 Rosewood Drive, Danvers, MA 01923, 978-750-8400. CCC is a not-for-profit organization that provides licenses and registration for a variety of users. For organizations that have been granted a photocopy license by the CCC, a separate system of payment has been arranged.

Trademark Notice: Product or corporate names may be trademarks or registered trademarks, and are used only for identification and explanation without intent to infringe.

Library of Congress Cataloging-in-Publication Data
Names: Dinero, Donald A., author.
Title: The TWI facilitator's guide : how to use the TWI programs successfully / Donald A. Dinero.
Description: Boca Raton, FL : CRC Press, 2017.
Identifiers: LCCN 2016011476 | ISBN 9781498754842 (pbk.)
Subjects: LCSH: Employees--Training of. | Employee training personnel. | Industrial efficiency. | Organizational effectiveness. | Teams in the workplace. | Industrial management.
Classification: LCC HF5549.5.T7 D5264 2017 | DDC 658.3/124--dc23
LC record available at https://lccn.loc.gov/2016011476

Visit the Taylor & Francis Web site at
http://www.taylorandfrancis.com

and the CRC Press Web site at
http://www.crcpress.com

Printed and bound in the United States of America by Publishers Graphics,
LLC on sustainably sourced paper.

To Maureen

My travel agent, muse, and life partner—You're always there to catch me when I fall. Three and I'm out.

Contents

Acknowledgments xv
Introduction xvii
Author xxv

1 A Brief History of Training Within Industry (TWI) 1
Training Within Industry Service 1940–1945 1
TWI from 1945 to 2015 7
Reintroduction 10
Summary 13
Notes 13

2 The Principles of TWI 15
Introduction 15
The Scientific Method 17
TWI Is Bottom Up (versus Top Down) 17
TWI's Main Objective Is to Solve a Problem and Not Just Deliver
Training 18
TWI Develops Confidence and Resourcefulness in How to Proceed,
Not Standardized Solutions and Rules 20
Use the Three J Programs Together, as Needed 20
TWI Is the Foundation of Lean, but Does Not Encompass All of
Lean 21
The Methods Should Be Applied according to the Stage of
Development of the Individuals and the Organization 21
The TWI J Programs Are Skill Based 23
A Key to Teaching Is Asking WHY 23
People Want to Be Productive and Be Involved in the
Organization's Operation 24

The TWI J Programs Help Satisfy People's Basic Human Needs 24
Notes 25

3 Understanding Job Instruction Training (JIT) 27
Why Use JIT 28
    Objectives of JIT 28
    Benefits of JIT 29
    Standard Work versus Standardized Work 30
    Creating Standard Work 31
Principles of JIT 32
    Learn by Doing: Everyone Must Do What It Is They Are Trying
    to Learn: Not Just Talk about It 32
    Use JIT When Training JIT 34
    One-to-One Instruction: Not Group 34
    Presenting Information in Small "Chunks" and Whole–Part–Whole ... 35
    Correct the Learner as Soon as He Makes a Mistake 37
    Repetition 37
    If the Person Hasn't Learned, the Instructor Hasn't Taught 38
    Break Down the Job: Use the Concepts of Advancing Steps and
    Key Points to Analyze, Understand, and Use the JIT Method 39
The "What," "How," and "Why" of JIT 39
    Preparation for Using the JIT Method 39
        Training Timetable 40
        Break Down the Job 41
        Prepare Everything 41
        Properly Arrange the Workplace 41
    The Job Breakdown Sheet 42
        Details: Identification and Order 42
        Order 43
    Advancing Steps and Key Points 44
    Terminology 47
    Creating a Job Breakdown Sheet 48
        Video 51
        The Heading 53
        The Body 55
        Recording Advancing Steps 55
        Recording Key Points 58
        Changing Key Points 60
        Software Applications 62

Training Aids 66
Critiquing a Job Breakdown Sheet 67
Finalizing a Job Breakdown Sheet 72
Get Consensus.. 72
Cascading 73
Delivering Instruction Using the JIT Method.. 79
JIT Four-Step Method.. 80
Step 1: Prepare the Worker.. 80
Step 2: Present the Operation.. 83
Step 3: Try Out Performance 88
Step 4: Follow-Up 90
Job Safety Training 90
JIT beyond Manufacturing... 92
JIT Creation Process.. 92
JIT Auditing and Updating... 93
Documentation... 95
Notes.. 96

4   Understanding JMT...99
Objectives of JMT.. 99
Benefits of JMT 100
Misconceptions of JMT 103
Continual Improvement Programs and Suggestion Programs............... 104
Principles of Job Methods Training... 106
1. Everyone Has Ideas.. 107
2. Most People Must Be Trained to Use Their Ideas 107
3. People Like to Share Their Ideas... 108
4. Break a Task into Small Details... 108
5. Question Every Detail: Question Everything........................ 108
6. Use the Six Questions, but in the Specific Order 109
7. Speak with Others... 109
8. Quantify 109
9. Get Approvals.. 110
10. Apply the New Method... 110
Using Job Methods Training.. 110
Describing How to Use JMT 114
Step 1: Break Down the Job 114
Step 2: Question Every Detail 115

Step 3: Develop the New Method 119
Step 4: Apply the New Method 122
The Proposal Sheet 122
   Heading 122
   Summary 122
   Results 123
   Content 125
Epilogue to the Viscosity Check Example: Making Changes 126
   Analysis Detail 7 126
Auditing 130
Documentation 130
Notes 130

5 Understanding Job Relations Training 133
Objectives of JRT 133
Benefits of JRT 134
Principles of Job Relations Training 135
Using JRT 139
   Define Your Objective 142
      Step 1: Get the Facts 142
      Step 2: Weigh and Decide 144
      Step 3: Take Action 145
      Step 4: Check Results 146
Changes to JRT 148
Auditing and Documentation 149
Notes 149

6 Selling (TWI) 151
Introduction 151
Standard Actions 152
   Step 1: The Initiator Researches and Talks about TWI 153
      Change 155
      Gap Analysis: Define and Quantify an Objective 155
   Step 2: Managers Accept a Small Pilot Program 157
      Meeting #1: Benefits 158
      Meeting #2: What Is TWI and Why Should We Use It? 158
      Meeting #3: Responsibilities and Commitment 163
Notes 166

7 Implementation 167
Preparation 167
  Finalize Plan 167
  General 167
  Job Selection and Training Room... 170
  Documentation, Assessments, and Publications 175
Training 176
  Points Common to All Three J Programs.................................... 176
Job Instruction Training.. 183
  Format.. 183
  Developing Coaches.. 184
  Developing JIT Trainers 186
  Training: Delivering the Sessions 188
    Day 1 189
      Step 1 193
      Step 2 194
      Step 3 194
      Step 4 194
    Day 2 195
      Item 1 (15 minutes: Open and Review)............................... 196
      Items 2 (20 minutes: Volunteer's Demonstration) and 3
      (5 minutes: Comments) 197
      Item 4 (15 minutes: AS and KP Concepts; JBS)................. 197
      Item 5 (40 minutes: Breakdown Knot) 199
      Item 6 (20 minutes: Breakdown Volunteer's Demonstration)........ 201
      Item 7 (5 minutes: Summary and Close)............................ 201
    Day 3 201
      Training Matrix..202
    Day 4 202
    Day 5 203
Instruction: Beginning the Use of JIT..204
Job Methods Training 206
  Format...206
  Coaching...207
  Developing JMT Trainers 207
  The Sessions...207
    Day 1 208
    Day 2 218

        Days 3–5 223
        Methods Improvement: Beginning the Use of JMT 224
    Job Relations Training 225
        Format 225
        Developing JRT Trainers 226
        Sessions 227
        Confidentiality 228
        Improved Relations: Beginning the Use of JRT 232
    Notes 232

8 Sustaining the TWI Programs 235
    The Ten Points 235
    The Kirkpatrick Model 239
    Embedding into a Culture 241
    Mandatory Actions for Success 244
    Notes 245

9 The Future of the TWI Programs 247
    A Look Back 247
    Beyond the Original Intent 249
    Creativity 249
    TWI as Foundational Skills 251
    Health and Personal Wellbeing 251
    The Learning Organization 253
    Coaching: BeLikeCoach 253
    Maslow's Hierarchy 258
    Human Performance Technology 259
    Risk Management 260
    Notes 261

Appendix I **263**

Appendix II **267**

Appendix III **273**

Appendix IV **279**

Appendix V **287**

Appendix VI **293**

Appendix VII **297**

Appendix VIII **299**

Index **301**

Acknowledgments

I began my TWI journey in 2002 and have been continually learning about its programs and underlying concepts ever since. I am grateful to my clients who continue to provide me with platforms on which I can delve more deeply into these programs in order to claim a greater understanding of them. The uniqueness of each organization has allowed me to see different facets of what outwardly appears to be a static, simple program. Because I have been privileged to see the TWI programs operate in a wide variety of environments, I have been able to question and refine my thoughts and understanding. One of my clients, Jaime Portillo of Toyoda Gosei Fluid Systems, was generous enough to share the Team Training Matrix he developed for his facility. I wanted to include it not only to show a contemporary training matrix, but also to demonstrate that a straightforward concept such as a training matrix can be developed into a sophisticated and useful management tool. Thank you Jaime.

I thank Mark Siwik, Swen Nater, Doug Saylor, and Alan Lambert of BeLikeCoach for helping me confirm that the TWI programs are not just for industry, but can be applied as well in athletics and other activities. John Wooden's coaching/teaching methods so closely align with those of TWI that it can be shown just how universal its principles are.

A special, heartfelt thanks goes to Dr. Louis Flaspohler, a physician at The Christ Hospital in Cincinnati. He called me one day, asking about the TWI programs and their applications in a medical environment because he was looking for any means through which he could improve the wellbeing of his patients in particular and people in general. He quickly saw how the TWI programs improve an organization by developing its employees and improving their wellbeing. Although he says he learns from me, I know that I learn much more from him. I am especially grateful for his detailed review of this manuscript and the excellent suggestions he offered.

Finally, I thank the editors of Taylor & Francis for giving me the opportunity to reach more readers with the powerful concepts of the TWI programs.

Introduction

Although the TWI programs were used extensively in the United States between 1940 and 1945, they fell into disuse by about 1970. They were later reintroduced into the United States in 2002. Since that reintroduction, many people have learned how to use them, but the results have been varied. In addition, not everyone who reads about TWI has the same amount of enthusiasm about these programs. For those of us who are very enthusiastic about TWI, we wonder why everyone does not see the benefits TWI offers as clearly as we do. This of course brings up the other question of "Why did people stop using these programs if they were so beneficial?" Answering the second question will help us answer the first. Answering these questions is important if, as we believe, the TWI programs benefit not only our organizations but also our society. It is well known that if we do not learn from the past, we will repeat mistakes previously made. The result is that our growth is slowed because we are learning through mistakes. It is much faster to learn from another person. Therefore, if we can determine how and why the programs fell into disuse in the United States about 1970, we should be able to make changes to prevent it from happening again.

A thorough research study that would answer these questions has not been conducted, but anecdotally there appear to be several contributing factors. Some people say that the veterans who returned to work at the end of WWII were already trained and so did not accept the TWI methods. That is questionable because the veterans were soldiers who were used to taking orders. Also, military training is similar to JIT and would not have been significantly new to them. Learning JMT and JRT would not have been a great leap. Additionally, many of those returning had never been in industry since they enlisted or were drafted right out of high school or college.

Another hypothesis is that once the federal government shut down the Service in 1945, there was no independent, recognized organization that had the authority to monitor and enforce quality and development. The material was in the public domain, and anyone who wanted to use or sell the TWI programs was allowed to do so. This seems to be a contributing factor since the TWI programs continued in countries when it was under government auspices such as Great Britain, New Zealand, and Japan. Furthermore, use of the programs declined in Great Britain and New Zealand when government support was withdrawn. Perhaps as important is the fact that when under federal control, the U.S. government paid for the training. After the war, companies had to pay consultants for what they used to get for free. If government control and/or oversight is a contributing factor, it appears that getting the federal government to regain control over the TWI programs is not a viable alternative. It is akin to putting the toothpaste back into the tube. That is, technically it is possible, but it would be considered a "Herculean task." An alternative would be for an independent organization to be recognized as a TWI authority. This seems unlikely today since all people offering TWI training are individual paid consultants and not a unified group.

A more viable hypothesis is that anyone who stops using the TWI programs simply does not understand them. That may sound difficult to believe and perhaps even arrogant, but although the programs are outwardly very simple, there is much to them. Understanding the programs is important because, much as we learn from our failures, we should also learn from our successes. If an organization's success is due at least partially to the TWI programs and this is not recognized, they may easily fall into disuse or be discarded outright. This concept of lack of understanding falls into two main categories. The first category refers to not knowing the impact that the programs have on the overall health of an organization. The second category refers to understanding the programs well enough to use them properly.

For the first category, let us look at a concept found in nature. In ecology, there is a concept known as a trophic cascade, where the elimination of a predator higher up the food chain can have significant results on the overall system. Two examples are the removal of wolves from Yellowstone Park and the over-hunting of whales. Whales, which have been hunted almost to extinction for their meat, bones, and oil, feed on fish and krill. When the whale population decreased due to over-hunting, the fish and krill population also decreased, which seems counterintuitive. One would think that when the whale population declined, the fish and krill populations would increase since a major predator is not as prevalent. However, it was then recognized that whales not only eat fish and krill, but they also sustain them. Whales eat fish and krill, and the fish and krill in turn eat plankton. Whales eat at lower depths where the fish and krill are, but they must rise to the surface for air. When they do that, they often release fecal plumes (excrement), which fertilizes the plankton. Fertilization of the plankton accelerates its growth so fish have more to eat and this, in turn, increases the fish population. In addition, since some plankton are plants, they give off oxygen and absorb carbon dioxide. When the amount of plankton increases, the absorption of carbon dioxide increases, which changes the climate. When the whale population decreased, the plankton was not fertilized as much and thus was reduced. This, in turn, reduced the fish and krill populations.

The TWI J programs, like the whales, are part of a larger system. Their intended purposes are to improve quality, productivity, safety, and cost, but when used properly, they go much beyond that. The programs improve communication, teamwork, and morale, but people may not recognize this. When people see TWI as only a series of "training programs" they easily dismiss them for a few reasons. The programs were introduced to improve training and quality, so if people believe they have a firm control of their training and quality, they may believe the programs are no longer necessary. What really happens is that when the TWI J programs are no longer used many other factors in the organization degrade. When the TWI J programs are no longer being used, the quality of communication will decrease. This has an influence on teamwork, perhaps creativity, and morale. The change will not be sudden, but it will occur. Since the programs also offer a standard method for three basic skills (instruction, improvements, relations), quality, productivity, safety, cost, and standardization will also slowly worsen.

The second category includes not understanding the programs and thus using the methods improperly. This does not seem likely at first glance because the programs are outwardly very simple. They are foundational programs and they can be applied to many different scenarios. If they contained much detail, their use would be restricted. The J programs were designed to be very practical so they could be used by anyone in almost any situation. As a result, they had to be standard. Each one has a four-step method, which makes it easy to learn and to remember. That enables their use, which results in success in meeting objectives.

The most widespread and obvious misuse of any of the J programs is the incorrect use of a Job Breakdown Sheet (JBS). In Job Instruction Training (JIT) many people do not understand the difference between what the Service called an Important Step and a Key Point. A very common mistake is to rewrite a Work Instruction or Standard Operating Procedure in three columns so that it appears to be a JBS. This lack of knowledge undoubtedly led to a reduction in the use of Job Breakdown Sheets (JBSs) because this change does not add value. The JBS is actually the "heart" of JIT and when it is not used properly, the performance of the JI method decreases substantially, although it might require an extended time period for that to happen. When the JBS is not written correctly, it cannot be used correctly. When the JBS is not used correctly, training degrades or is no better than any other training program. Since JIT affects other aspects of an organization such as communication and teamwork, improper use of JIT can have a deleterious effect on not only those factors but also JMT and JRT.

The JBS is structured to tell us WHAT to do in a job, HOW to do it, and WHY we should do it that way. The same is true of learning anything, and that includes learning the TWI programs. We first have to know WHAT to do and then we learn HOW to do it so that the final result is as designed. Understanding WHY we do something is not required if there are no problems and everything occurs as it should. However, if results do not occur as intended, one of three actions is usually taken.

1. They may stop using TWI altogether (since it doesn't work).
2. They may seek help from someone who knows how to use it properly.
3. They change the method to "improve" it.

Option 1 answers the question directly. When people do not use the method correctly, it does not work and they assume TWI does not work in their particular organization. They do not know that when used properly, the TWI programs work in any organization because they are based on human characteristics and needs. Option 3 occurs because many see the programs as simple devices that were created 70 years ago and need "upgrading." Without knowing and understanding the basic principles, changes are made that make the programs less effective and then they revert to Option 1. What also happens is that the programs are made more complicated and thus are limited to fewer people, again making them less effective. Option 2 is the correct one, but a person must be available who knows the TWI principles well enough not to make mistakes as in Option 3.

Because people may not have known why they were doing aspects of the J programs, they may have stopped doing those aspects. Today, knowing why is more important than ever because matters are changing faster than ever. For example, today some people believe they can become competent in any of the J programs in a day or so. What they do not recognize is that although they may have some knowledge of the subject, they cannot use it correctly.

Knowing why we do actions is important because our environment is constantly changing. Our society is different from what it was 75 years ago. Our needs and our organizations' needs are different, so the J programs must also be different to meet those needs. Changing the manual from the vernacular used in the 1940s to that used in 2015 is a change that should be made, but it is superficial. Changes must be made that reflect cultural changes. However, we must maintain the principles used in the programs while we change the delivery of the programs to suit the culture. The principles are those ideas that are characteristic of human behavior. We can see that they are as pertinent now as they were 75 years ago and will remain relevant until human behavior and capabilities change. Therefore, before we make any changes, we must understand what the TWI principles are. If we do not, then we will weaken the programs and they will again fall into disuse.

The purpose of this book therefore is to prevent the TWI programs from falling into disuse as they did in the United States 40–45 years ago. At the same time, the 3 "J"s should be engrained in our society as much as the 3 "R"s. The principles of these programs are as valid today as they were 75 years ago because people have not changed significantly since then. I will attempt to keep the TWI programs alive and well by

- Clarifying misunderstandings and misconceptions
- Telling WHY we do what we do when we use TWI
- Explaining the basic TWI principles
- Maintaining their simplicity

This book will explain what the TWI programs are; However, it is also important to know what the TWI programs are not. They are not higher-level programs like Lean/Six Sigma, which are meant to address and solve all or most of an organization's problems. They are not a "cure-all" or a "silver bullet" that solves all problems. Although they can be modified, the main principles must not be changed until the human condition changes. They cannot be mastered in a few hours or even a few days. Most people take a few weeks with much repetition and many examples. You cannot become skilled at these programs just by reading about them. You must actually perform the skill to master it. The repetitive performance must be accompanied by correct external feedback (a coach) if you want to master the skill in the shortest time. Just as mastering the 3 "R"s is a life-long process, mastering the 3 "J"s will take a similar amount of time.

Notes

1. Gino, F. and Pisano, G. "Why Leaders Don't Learn from Success." *Harvard Business Review*, April 2011. https://hbr.org/2011/04/why-leaders-dont-learn-from-success.
2. Plankton are microscopic organisms. Phytoplankton are classified as plants and zooplankton are classified as animals.

Dictionary of Acronyms

BPNT   Basic Psychological Needs Theory
JBS   Job Breakdown Sheet (used in JIT)
JIT   Job Instruction Training
JMT   Job Methods Training
JRT   Job Relations Training
MBS   Job Methods Breakdown Sheet (used in JMT)
PD   Program Development
SDT   Self-Determination Theory

Definitions of Positions

- Participant: A person who attends a 10-hour J program
  - This person knows the respective J program, but she or he is not particularly good at using it. They understand the concepts and the mechanics but they need more practice under the supervision of a coach to strengthen the skill.
- Instructor: A person who has completed the 10-hour JIT program
  - This person is a JIT participant who has used the JIT method enough that she or he can create a correct draft JBS, help get consensus on it and correctly deliver it to individuals through instruction.
- Coach: An instructor who has enough experience creating and using JBSs that she or he can assist participants in developing their skills. This would also apply to JMT and JRT.
  - Although there is no set number of examples that must be experienced to become a coach, most people are comfortable after 15–20 repetitions. This person must gain much of this experience under the supervision of an Institute Conductor or trainer so that she or he gains not only the knowledge of the program but also the methods and questions to ask when interacting with the participant.
- Trainer: A coach who can also deliver the 10-hour J program
  - As with any skill, using the program and instructing someone else in how to use it are separate skills. Coaching is used during the 10-hour program, which is why any trainer must first be a coach.

- Institute Conductor: A trainer who can develop others to become trainers. This usually requires delivering the given program at least 6–10 times.
  - Much of the work and time required to become a TWI trainer comes from the trainer candidate him- or herself. However, there are some key points that an Institute Conductor can offer which will shorten the learning curve.

Author

Donald A. Dinero, PE, CPIM, has almost 50 years of experience designing and implementing manufacturing methods and processes, and is the principal of TWI Learning Partnership. He has addressed problems in all aspects of operations including but not limited to change management, personnel, labor unions, production systems, and production control.

His BS degree in mechanical engineering is from the University of Rochester and his MBA and MS (career and human resource development) degrees are from the Rochester Institute of Technology. Don deliberately sought degrees in these areas so that he would have a balanced academic background in technology, business, and organization development. He received his Professional Engineering license (NYS) in 1983 and his Certification in Production and Inventory Control from APICS in 1986.

After more than 30 years in positions of manufacturing and engineering management and as a direct contributor, Don entered the Lean consulting field by joining existing consulting firms. In 2002, he learned about TWI (Training within Industry) and its reemergence in the United States. After being shown the three J courses, he began to study all the materials he could find on the subject. His studies and talks on TWI led to his writing the book *Training Within Industry: The Foundation of Lean*, published by Productivity Press, 2005. This book won a Shingo Prize for Research in 2006. *TWI Case Studies: Standard Work, Continuous Improvement and Teamwork,* also published by Productivity Press, followed in 2011.

He recognized early on that the Lean movement is hindered by its omission of TWI. TWI offers fundamental skills training, which helps to stabilize an organization, preparing it to seriously begin its Lean journey. In addition, it provides a foundation so that Lean principles are sustained. In order to assist in stabilizing an organization and thus assist in the acceptance of Lean, Don concentrates his efforts on spreading the word of TWI. His consulting practice focuses solely on the TWI programs. To that end, he delivers training in all three of the J programs and in Program Development. In keeping with the "multiplier effect" cited by the Training within Industry Service, Don also offers Train the Trainer sessions for all four programs. This allows an organization's employees to independently deliver the training.

As Don continued to attend to the needs of clients and spread the word of TWI through conferences and articles, he became aware of two ideas. First, many people did not truly understand how to use the TWI J programs and second, that most people did not recognize that the most important benefits of TWI lay beyond the initial gains. The TWI programs are not just a foundational element of Lean thinking, but of several other significant bodies of knowledge such as Learning Organizations and leadership, for example.

Knowing that the TWI programs consist of fundamental pedagogical principles, Don persists in spreading the word of TWI to all organizations, knowing that any organization can benefit from these programs in many ways.

As a student of TWI, Don continues to have a desire to learn and improve with respect to the TWI programs. He welcomes all input and feedback and can be contacted through email at dadinero@TWILearningPartnership.com

Chapter 1

A Brief History of Training Within Industry (TWI)

> Those who cannot remember the past are condemned to repeat it.
>
> **George Santayana**

Training Within Industry Service 1940–1945

In 1939, much of the world was at war. Adolf Hitler's forces were about to occupy all of Europe, and Japan had occupied part of China and was at war with the Soviet Union. Although Germany was sinking U.S. freighters along the eastern coast of the United States, there was a preponderance of isolationists preventing the U.S. Congress from declaring war on any country. The feeling was that World War I was "the war to end all wars," and this war had nothing to do with the USA. Franklin Roosevelt, the president, did not agree, but since he did not have authority to declare war, he used other methods to oppose the isolationists and help the Allies. In 1939, the Allies were Great Britain, France, and Poland. By the time Roosevelt had taken action, France and Poland were occupied by Germany, and the Allies consisted of the British Commonwealth. It was at the end of 1940 that Roosevelt had come up with a plan to arm and support the Allies. In a radio broadcast, he referred to this as "The Arsenal of Democracy." America would supply armaments to the Allies, but would stay out of the actual fighting.

One main problem was that the U.S. industrial sector was not in a position to supply Britain with what it needed. Unemployment was about 25%

and capacities were far from where they should be for such a commitment. The government knew changes had to be made quickly and productivity had to increase dramatically. One of the many efforts made was the creation of the Training Within Industry Service.

> Training Within Industry was an emergency service to the nation's war contractors and essential services. Its staff was drawn from industry to give assistance to industry, and its history covers the time from the Fall of France to the end of World War II—from the summer of 1940 to the fall of 1945. TWI's objectives were to help contractors to get out better war production, faster, so that the war might be shortened, and to help industry to lower the cost of war materials.[1]
>
> It is not possible to try to understand this World War II agency called the "Training Within Industry Service" without looking at the backgrounds of the four men who developed and directed it: C.R. Dooley, director; Walter Dietz, associate director; M.J. Kane and William Conover, assistant directors. They had known each other for years and shared the same philosophy of training for production, although each brought with him to TWI his own special experience and talent. Each joined TWI in 1940 on loan from his employer without government compensation.
>
> Mr. Dooley had three industrial connections—with Westinghouse Electric and Manufacturing Company which he joined in 1902, with the Standard Oil Company of New Jersey, and with the Socony-Vacuum Oil Company whose industrial relations manager he was when he came to Washington in 1940. In all of these companies the planning and direction of training was part of his responsibilities.
>
> Mr. Dietz joined the Western Electric Company in 1902 and has been continuously associated with that company except when he has been on loan to the government for wartime assignments. He held the position of personnel relations manager of the Manufacturing Department when he came to Washington in 1940. Both Mr. Dooley and Mr. Dietz remained with TWI throughout its five years of operation.
>
> Mr. Kane had been with the General Electric Company as a personnel manager before the first World War, and after that war went to the American Telephone and Telegraph Company where he was staff engineer on training of supervisors, instructors, and conference leaders when he came to TWI in 1940. After spending almost four and a half

years with TWI, Mr. Kane left on January 1, 1945, to become director of industrial relations for the National Association of Manufacturers.

Mr. Conover came to TWI in 1940 from the United States Steel Corporation where he was assistant director of industrial relations. His previous industrial connections were with the Philadelphia Gas Company, the Western Electric Company, and Lycoming Manufacturing Company. He left TWI in December 1944 to join General Cable Corporation where he is director of manufacturing.[2]

Much more has been written about the TWI Service,[3] but suffice it to say here that the efforts of this service were extremely successful and they have been referred to as "the most underrated achievement of 20th century industry."[4] It is the only government agency to be given an award by industry for its service to industry. The men driving the Service, named above, were experienced managers who knew how to increase production, but their secret was in knowing something that many contemporary managers miss.

> If there is any single thing that could be stated as "what TWI has learned" it would be that the establishment of principles, and even getting acceptance by managers, alone have practically no value in increasing production. What to do is not enough. It is only when people are drilled in *how* to do it that action results.[5]

Based on their experience, these men believed that at least 80% of errors occur because people have received no training or poor training. Their underlying belief was that people want to be productive and thus, if they know what to do, they will do it. (Refer to Appendix VI—The Philosophy behind TWI.) The main reason people do not do what is expected is because they have not been trained properly. Most training occurs through trial and error. When we learn through trial and error, most learning takes place once we have made an error. Consequently, time, scrap, rework, and injuries all have a tendency to increase. Aside from the monetary costs involved, the country was considered to be in "an emergency situation" and could not afford the lengthy time that trial and error training takes.

The service faced two main problems as they began. What information should be given to the contractors to increase their productivity and how should that information be disseminated? In order to answer the first question, they asked the defense contractors what they needed. They did this initially and throughout the five years of their existence. They would attempt to meet the needs of every contractor they encountered. Before they even got a chance to set up their offices, they became aware of a need "for 350 qualified lens grinders for work in government arsenals and navy yards [since] the Employment Service and Civil Service failed to locate skilled men."[6] Training was obviously a need and at least one function that should be delivered. At this time, a lens grinder apprenticeship was about five years, so it was obvious that this need had to be satisfied some other way.

Since they did not have an appreciable budget (their salaries were paid by their companies), they could not hire many people. Although many served as they did without government compensation, it was necessary to pay some employees. With insufficient personnel to cover the entire country, the question was how to spread knowledge in a timely manner.[7] Over the course of the next year, they tried sending consultants to companies to help. There were two main difficulties with this. The first problem was that there were many more defense contractors needing help than there were available consultants. The second problem was that once a consultant left the facility, operations would return to the original plan used before the consultant arrived. Any gains made were lost.

The next attempt was to record what was to be done and send this information to contractors in the form of bulletins. Again, this met with two difficulties. The larger problem was that the only people who would read the bulletins were those in libraries or schools. If someone in a manufacturing facility did read the bulletin, he or she was not given enough detail to be properly trained for their specific situation.

The need for fast, reliable instruction was satisfied as they developed the Job Instruction Training (JIT) program. Using the first iteration of this program, which was mainly gained from experience in World War I, skilled lens grinders were developed in 5 months as opposed to five years. By 1945, the time was reduced to 6 weeks.[8] The need for a dissemination of the training was addressed by what they called "the multiplier effect." A TWI representative would enter a facility (on request only) and would train employees to use the JIT method. One or more of these employees would be chosen to become JIT trainers and they would receive additional training in how to deliver the 10-hour program. Once that had been accomplished, the representative would no longer be necessary and he could attend another facility. The thinking was that training should occur "within industry by industry."[9]

As the TWI Service reached out to more and more companies, they continuously asked for the needs of individual contractors. Interestingly, as they gathered data, they began to realize that all defense contractors (all companies) had the same basic needs. That should not be too surprising since they were all trying to make a product and they were all trying to do it with people. As they collected the data, they found that they could categorize it into five basic needs. They referred to this as the Five Needs of a Supervisor. They used the term *supervisor* because, lacking both time and money, they knew that they could not reach every member of an organization. They thus wanted to concentrate on the first-line supervisors who acted as a central point for production. They were at the intersection of the production worker and management, so by training first-line supervisors, they could effect the largest change with minimal effort. They did note that if time permitted, other personnel should be involved. As a result, today we refer to the Five Needs of an Employee since everyone has these needs and everyone can benefit from satisfying them. The underlying premise is that everyone has these needs and if people can master them, they will be successful in whatever they do. The five needs consist of two categories: knowledge and skills. This is an important dichotomy to understand because we can instruct people to learn a skill but they must absorb knowledge by themselves. Technical terms, for example, are representative of knowledge. We can only tell someone what they are and s/he must understand and remember them. Making a product is a skill, which can be instructed, and the use of technical terms will help in that instruction. Differentiating between knowledge and skills works on several levels and helps us understand and use the JIT method. The five needs are

1. Knowledge of work: This has to do with the technical aspects of your job. If you are an accountant, it would be the accounting system you use, and so on. If you are a welder, it would have to do with knowledge of the equipment and the metals, and how to join them.
2. Knowledge of responsibilities: This has to do with the policies and procedures of the job. It could include benefits, safety clothing required, or relationships with other departments or people.
3. Skill in instructing: This reflects the transfer of knowledge to another person. It could be formal training or informally just giving someone an assignment.
4. Skill in methods improvement: This refers to improving what you are doing. Everybody has ideas on how to do their job differently, but not everyone knows how to vet, sell, or implement his or her ideas.

5. Skill in relations: Everyone deals with other people to some extent, but not everyone knows how to build strong positive relationships that are necessary for work to be as effective as it can be. When personnel problems do arise, everyone should know how to successfully solve them.

Because of their expertise and experience, the four TWI developers believed that good training was the basis for improving productivity. Furthermore, they believed that the multiplier effect was the proper vehicle to disseminate the knowledge to all contractors. In order to do that, however, they would have to create programs that could be easily learned by almost anyone. They knew each contractor would not have a professional trainer available to deliver the program, so they had to create programs that anyone could easily learn how to deliver. As they looked at the five needs, they had developed such a program for instruction, and knew they could create similar programs for methods and relations. Each program took several iterations until they were satisfied enough to release it for actual use. They were most satisfied with Job Instruction because they had worked on that the longest. When the Service was closed in 1945, they knew that each of the programs could be improved or at least changed. They knew changes were necessary because these programs served the needs of people and they knew people's needs change. Although each of the programs has changed somewhat over the years, the core programs remain the same.

One reason for this has to do with the Self-Determination Theory (SDT). SDT is a theory of human motivation and personality. It was developed in the 1970s and was accepted as a sound empirical theory in the 1980s.[10] As a theory of motivation, it deals with why people do what they do. It states that every human has three basic needs. We are all trying to satisfy these needs in whatever we do.

> Conditions supporting the individual's experience of *autonomy*, *competence*, and *relatedness* are argued to foster the most volitional and high quality forms of motivation and engagement for activities, including enhanced performance, persistence, and creativity.[11]

In other words, in everything we do, we are seeking a degree of autonomy. That is, we want to have some control over what we are doing. We also seek competence because we want to know how to do what it is we are doing. Having to do something that you do not know how to do can be very frustrating and thus irritating. Since we are social beings, we also desire to get along with those around us. Autonomy is addressed by Job Methods because when we use Job Methods Training (JMT), we are having an impact on how we do our jobs. Someone gives the job to us, but when we use JMT, we have an influence on how we do it and thus have a degree of autonomy. Job Instruction addresses competence since it teaches us everything we need to do the job as well as it has to be done. The result is that we become a skilled operator in that job and possess all the necessary competence. Job Relations teaches us how to develop strong, positive personal relationships and thus address the relatedness factor. Because the three J programs relate so closely to SDT, it leads one to believe why most people are so receptive to them. It also explains that the five needs are not just arbitrary, but are based on human behavior.

In the Five Needs of a Supervisor, the first two needs are Knowledge of Work and Knowledge of Responsibilities. These two knowledge categories varied so much that they could not create a standard program for them. The knowledge of work and knowledge of responsibilities varies from company to company, and from department to department. It even varies within a department. Situations related to these needs did arise and interfered with productivity, but it was not possible to create a standard program to cover every possible one of them. What they could do, however, was to teach people how to solve problems that were specific to a given situation. The result was the creation of Program Development, which taught people how to analyze a situation and develop a solution to problems that were recognized. All problems are either man-made or machine-made. If they are machine-made, we apply Job Methods to improve equipment, a procedure, or a system. If the problem is man-made, we apply Job Instruction to improve a person's skill, or Job Relations to improve a person's attitude or behavior.

TWI from 1945 to 2015

The TWI Service was notified in May 1945 that it would be closed in September of that year. Although the war was not officially over, it was a foregone conclusion that the Allies would be victorious. Since the Service was created for the war effort, it seemed logical that it should be disbanded at the conclusion of that conflict. The directors disagreed, but to no effect. In a letter of transmittal of the "Training within industry report, 1940–1945," C.R. Dooley wrote

> We have learned so much about the techniques of training that what we knew before is as nothing. This learning has been at the expense of the taxpayers and therefore should be preserved and used in peacetime. These techniques are as applicable to peace as to war production.[12]

The TWI Service treated these programs as a great laboratory experiment. Their underlying thinking was based on the Scientific Method. In fact, the developers wrote about the need for a problem-solving program based on the Scientific Method (analyze, plan, execute, evaluate).[13] During its 5 years in existence, the J programs, as they were known, were constantly refined. They also knew that the programs should continue to change, and wrote in their report:

> Development work would have continued as long as TWI existed—no program is ever perfect, and no program is any good unless it meets needs. Since needs change, any program must be kept growing.[14]

Although the Service was disbanded in 1945, millions of people in the USA, Canada, Great Britain, and Australia had benefited from the programs. In addition, many of these people not only knew how to use the programs, but also knew how to train others to use them. In 1946 in the USA, the four men who were the main drivers of the TWI Service incorporated an organization called The TWI Foundation.[15] It was an association of member companies and acted much like the TWI Service did. That is, it continued to develop the TWI Programs and even added a new one in economics. Many of the other men in the TWI Service formed consulting companies. Most notable was Lowell A. Mello, of Cleveland who formed T.W.I., Inc.[16] This company was notable because it won the contract to introduce TWI to Japan in 1950. However, other consultants continued to spread the TWI programs to countries around the world, including

- Italy
- France
- Spain
- Belgium
- The Netherlands
- Luxembourg

- Turkey
- Indonesia
- New Zealand
- Ireland
- Nepal
- India
- Vietnam

among many others.[17] The book that Walter Dietz published in 1970, *Learn By Doing—The Story of Training Within Industry 1940–1970*, makes it sound like the programs were flourishing around the world in 1970. If that is the case, in 1970 they were at the peak of their existence. Little is found in the USA after 1970. As good as these programs were, they slowly fell into disuse. In America, where they developed over a million workers, little evidence can be found of their use after the early 1970s. This author worked at Kodak in Rochester, NY, in the 1990s and, like an archeologist, saw evidence of their existence at one of the companies responsible for the historic "lens grinder study."[18] At that time, I could find no one who knew what an *Important Step* or a *Key Point* was, yet it was obvious that there was space available for them in the operator's work instructions that were still in use. Since Rochester was the *home* of Kodak, I had occasion to speak with many retirees who still live in Rochester. One was associated with training and had a copy of a manual titled "Job Instruction Training," which was copyrighted in 1968 and revised in 1969. The intent of the manual was to be a self-instruction manual. If a person could learn JIT by himself or herself, there would be no need for an instructor and costs could be cut. This plan did not work out very well and this individual was asked to rewrite the manual to make it more usable. His manual stated, "This is a teach-yourself book for people who teach people. If you have to teach someone else how to do a job, this is for you. Turn the page and start the fun."[19] Although this manual was done very well, it missed one of the TWI principles that it is very difficult to teach something as simple as JIT to yourself. Isaac Newton and Gottfried Leibniz created calculus,[20] and some people learn it by themselves. Most people, however, require a teacher to learn calculus. In fact, calculus is more suited to being self-taught because answers to problems are either right or wrong and one can find enough examples in textbooks to drill sufficiently to learn the method. The same is not true with the TWI programs since there is no right or wrong Job Breakdown Sheet or resolution to a Job Relations situation. Consequently, the effort to continue the use of JIT slowed and by the 1990s had stopped entirely.

The countries that put the TWI programs under federal government control seem to keep using them the longest. We have a letter from a representative of the British Ministry of Labour to the Industrial Training Service in New Zealand. The representative thanks the recipient for the material he received and bemoans the fact that "the United Kingdom TWI side seems to be winding down." The letter is dated 1982. New Zealand followed suit about a decade later. Of course, just because the government closes its training bureau does not mean the TWI effort stops. Consultants in New Zealand are still using the programs. The disadvantage of not having government involvement in these programs is that there is no central body to continue their development or maintain their quality. Toyota started using the TWI programs in 1951 and has continued to use its principles ever since. Today, 70 years later, one might not recognize the TWI programs at Toyota, but the principles may still be used.

Reintroduction

Although the use of the TWI programs had faded in the USA by 1970 or so, they did not disappear completely. Practitioners such as Jim Huntzinger[21] and academicians such as Robinson and Schroeder[22] wrote articles about them. The U.S. Department of Labor, Employment and Training Administration included "A step-by-step process to train an older worker" on a page in their website,[23] which is no longer active. In 2001, Bob Wrona, a project manager at the Central New York MEP site (Manufacturing Extension Partnership—aka Technical Development Organization—TDO) saw the letters "TWI" as a footnote in a book and began an investigation as to what it was. Once he found out, he saw the value in it and began a campaign to reintroduce the programs into America's industry. The programs were delivered much as they had been in the 1940s, and where they were used properly, the results were similar. Companies that used JIT by training employees and getting many people involved expended some effort, and found that the rewards overshadowed the costs. The fact that this campaign was not expanding as fast as some thought it should may be accounted for by a general skepticism based on false promises of "programs of the month" that many people experience. Initially, there was only data that had been created during World War II, but that was often discounted because it was old and the situation was different when it was collected. As contemporary data started to accumulate, the pace of acceptance still did not seem to increase. The nagging question remained: "If these programs are so good, why did we stop using them?"

There were two related factors that may directly affect the contemporary life of the TWI Programs. First is the fact that with this "reintroduction" in 2002, no organization equivalent to the TWI Service was created. There is no central, independent organization that would monitor the quality of the programs and continue to develop them as the original developers had hoped. In our competitive society, people learn about the TWI programs, access the manuals, and sell themselves as "TWI Training Consultants." In some cases, this turns out well, while in others TWI becomes another "program of the month."

The second factor is how training is thought of today. The TWI Service had to convince management that training was one of its responsibilities. It was no less necessary than equipment or raw materials in a production environment, which includes a shop, or a lawyer's or accountant's office. Today, it is more common to have a training department; but even so, many of today's managers believe that training is someone else's responsibility. In some companies, production supervisors relegate all training responsibilities and execution to the training manager. Although the execution may be delegated to a training manager, the responsibility remains in production. Training should be a high management priority and if top management does not play a major role in its implementation and acceptance, it will fail. A common refrain is that managers cannot find enough skilled workers. Instead of asking for skilled workers, the request should be for workers who can be trained to be skilled. Adding to this problem is the plethora of programs from which managers can choose to achieve their goals. Most are good but no program will add value if it is not used properly. If training were thought of as a problem-solving tool, it would be much easier to separate good training programs from others. So, in many of today's organizations training is thought of as a "necessary evil" at worst, or "something nice to do when there is time" at best. There are, however, some organizations that take training seriously and consider it integral to the production process. When the TWI Programs are presented to them as they were in 1940, the programs may seem good but not great. A possible reason for this is that the TWI Programs are often presented as WHAT to do. People are often looking for a recipe or a standard procedure. Although the TWI Service made a point of simplifying the methods so each J program used a four-step method, they intended those to be a framework to be adapted to an individual situation. They wrote:

> Confidence and resourcefulness in how to proceed, not standardized solutions and rules, are developed.[24]

For example, the JIT trainer's manual says to perform the job three times for the learner. The first time the instructor should just tell him the steps. The second time the job is done, the instructor should say the steps and any key points. During the third repetition, the reasons should be added. Questions often asked are, "Do I have to do the job three times?" "What do I do if he doesn't understand it after three repetitions?" The point is that the instructor's goal is to transfer given information to the learner and the three-iteration process is generally recommended. Similarly, the instructor should watch the learner perform the job four times.[25] However, the instructor must use his judgment as to how much he should deviate from the standard method to accomplish his objective. There must be enough repetitions to show the instructor that the learner both understands and remembers the task. If a task is very simple, understanding might not be a problem but remembering may be. This often happens when instructing a job on a computer. The keystrokes are simple, but remembering what they are can be challenging. When this concept of having flexibility in the number of iterations is not presented to the instructor, he may think that JIT is too rigid for his use and will not work in his organization. This is why knowing the principles of TWI is very important.

The basic principles of TWI are based on human characteristics and will be applicable until humans evolve beyond where we are now. We put on a coat to go outside when it is cold. We did this 200 years ago and we will do it 200 years into the future. The coats will look different, but they will still be coats. The TWI Service treated these programs as a great laboratory experiment. Their underlying thinking was based on the Scientific Method. In fact, the developers wrote about the need for a problem-solving program based on the scientific method (analyze, plan, execute, evaluate).[26] During its five years in existence, the J programs were constantly refined. The TWI Service also knew that the programs should continue to change as noted previously.It is unknown by this author how much development work continued in various countries.

The TWI Programs proved to be extremely successful and are based on the characteristics and behaviors of the human condition. Yet, we see that they fell into disuse. Now that they have been reintroduced, we must make sure that the elements that add value are not lost again. Having a better understanding of these programs—WHAT they actually consist of, HOW to correctly use them, and WHY we should use them in a certain way—will increase the likelihood of their retention. Let us take a deeper look into TWI and find the answers to those questions.

Summary

- The TWI Service was created to increase productivity among all U.S. defense contractors during World War II.
- It has been referred to as "the most underrated achievement of the twentieth century."
- The solution to increasing productivity lay in improving training, and that should be done "within industry by industry."
- The programs are relevant today because they are based on human behaviors and needs (autonomy, competence, relatedness).
- Companies that achieved gains by using TWI no longer use the programs.
- Understanding the principles of TWI will prevent us from incorrectly changing it, which will lead to its continued use.

Notes

1. Training Within Industry Service. 1945. "Training within industry report, 1940–1945." Washington, DC: War Manpower Commission Bureau of Training, preface.
2. TWI report, Foreword.
3. Dietz, Walter. 1970. *Learn by Doing: The Story of Training within Industry*, Summit N.J.: Walter Dietz.
4. Dinero, Donald A. 2005. *Training within Industry: The Foundation of Lean*. Boca Raton, FL: CRC Press, p. xi.
5. TWI report, Preface. The framework of "What," "How," "Why," which is plainly stated in JIT, is a powerful concept that works on many levels and is critical in learning theory.
6. TWI report, A Study of Lens Grinding, p. 271.
7. TWI report. The most personnel at any time (paid or unpaid) was 410 for the entire country, p. 310.
8. TWI report, chapter 2, p. 20.

9. This is a brief overview, and additional information can be found in *Training within Industry: The Foundation of Lean*, CRC Press, Donald A. Dinero, 2005. Original documents can be found on www.trainingwithinindustry.net.
10. https://en.wikipedia.org/wiki/Self-determination_theory.
11. www.selfdeterminationtheory.org/theory.
12. TWI report. Letters of Transmittal.
13. TWI report, p. 267.
14. TWI report, p. 261.
15. *Learn by Doing*, p. 7.
16. *Learn by Doing*, p. 110.
17. *Learn by Doing*, p. 64.
18. *Training within Industry*, p. 92.
19. Factory J.I.T., Young, David R., Kodak Apparatus Division Training Department, 1977.
20. *A Brief History of Calculus*, https://www.wyzant.com/resources/lessons/math/calculus/introduction/history_of_calculus.
21. Huntzinger, Jim. 2005. "The roots of Lean: Training within Industry and the origin of Japanese management and Kaizen." Lean Enterprise Institute.
22. Robinson, Alan G. and Schroeder, Dean M. "Training, continuous improvement, and human relations: The U.S. TWI Programs and the Japanese management style." *California Management Review*, 1993, Vol. 35, No. 2.
23. www.doleta.gov/Seniors/html_docs/docs/unique1.cfm.
24. TWI report, p. 35.
25. First to see the Learner do the job, second to hear Advancing Steps, third to hear Key Points and fourth to hear Reasons.
26. TWI report, p. 267.

Chapter 2
The Principles of TWI

> The man who grasps principles can successfully select his own methods. The man who tries methods, ignoring principles, is sure to have trouble.
>
> **Harrington Emerson**

Introduction

People and organizations vary over time and also with respect to other people and companies. Therefore, a given activity may be very successful at one point in time for one individual or one company, but that same activity may be less effective at another point in time or if it is transferred to another person or company. A principle, however, is much more stable. A principle is defined as "a fundamental truth or proposition that serves as the foundation for a system of belief or behavior or for a chain or reasoning."[1] Some principles are culturally or technically derived and thus can change, although they are not likely to. Examples of this can be found in medicine. For years it was thought that if a patient were ill, removing some of his or her blood would aid in the cure. If a doctor did not perform bloodletting on a sick patient, the doctor could be accused of malpractice. Today, we know that bloodletting usually harms the patient. Some principles describe natural phenomena and will not change. Gravity is one example that functions the same way throughout the universe. For our purposes, it is not necessary to differentiate between the two types. It is only necessary to recognize that there are some principles in what we do. It is important to know and understand the principles of the Training within Industry (TWI) programs because we are constantly experiencing changes. We change jobs within an organization, we change companies, and we change our behaviors. We also change as a society, both in general and where we live and work. The changes may be abrupt or they may be subtle and we might not even notice them. Although changes are happening all around us and, indeed to us, the principles we use remain stable. If we change or ignore the principles of our system when we respond to a situation, the result may not be what we intended.

When we perform an activity, it is composed of principles and other actions that are suited to the particular situation. The TWI Service was charged with increasing the productivity of all the defense contractors in the USA during World War II. The developers were aware of the principles required to accomplish this goal, but they had to determine methods so that those principles would be followed across the country. They created four programs based on these principles and on the culture in the country at the time. They knew that cultures varied from company to company and so they had to create programs that were blind to these differences. Although people have not really changed much since TWI was developed in the 1940s, society has. The TWI Service could not have created methods to be used in today's society because they could not see today's society. If they had, the 2015 methods would not have been as effective in a 1940s' society. For example, the Trainer's Manual says we should put the participants at ease at the start of a session because people learn better when they are not nervous. The manual suggests lighting a cigarette. That may have been acceptable in 1940, but it would not be acceptable today. What has not changed, however, is that the participants should be put at ease and made to feel comfortable because it still holds true that people learn better when they are not nervous.

The men of the TWI Service recognized two fundamental ideas about the programs they created. First, they believed the programs, as written in 1945, could be improved. Job Instruction Training (JIT) was mature since they had been working on it for four years, but the other two J programs, especially Job Methods Training (JMT), had potential improvements about which they were thinking when the Service was closed. Had the Service not closed in 1945, these two programs most likely would have been revised in 1946. In fact, the organizations that took over the main TWI effort after 1945 did revise these programs. The second fundamental idea that they recognized was that they had created these programs for a 1940s' society. The programs met the needs of that society, but they knew that the needs of a society change. Thus, the programs must change along with it. The changes would include aspects of the methods, but the principles should remain the same. The principles were based on human behaviors and should change only when basic human behavior changes. It is important to read what these men wrote because it tells us not only what they were thinking, but also how they thought. Knowing what and how they thought leads us to understand the main concepts and principles they used in creating these powerful, yet simple, programs.

The TWI programs contain some principles that are common to each individual program and others that apply only to each specific program. These principles are the ideas that should not be changed. They are universal and apply to all people. They are as true now as they were in 1940. When these principles are altered or ignored, the effectiveness of the programs diminishes. Before we review these main principles of each of the individual TWI J programs, let us review the principles that apply collectively to all the programs. The principles are important because we will often be confronted with a situation that we have not seen before. We must be able to stop and refocus to remember what we are trying to accomplish and how we plan to get there. These principles form the basis of the TWI Programs.

The Scientific Method

The Scientific Method, which has been used for over 300 years, is a technique used for answering a question. The number of steps varies with the writer, but the main scheme is to observe a problem, hypothesize a solution, act, review results, and repeat as necessary. This thinking is fully embedded in the J programs. It is most notable in Job Relations Training (JRT), where the four-step method mirrors the steps in the scientific method. However, it is also in JMT, which emphasizes asking "why" and has as an objective developing a questioning attitude. It is, perhaps, most subtle in JIT, where we are taught to question everything we do and must provide reasons for all actions that do not advance the work.

TWI Is Bottom Up (versus Top Down)

When the TWI Service began designing its programs, it focused on the first-line supervisor. That would be the person in management who is closest to where the value is added to the product. These supervisors alone were meant to get the 10-hour training and drive the programs because the country was in an "emergency situation." That meant they could afford neither the time nor the manpower to train everyone to deliver TWI. If resources permitted it, they encouraged every person to participate in the training because they knew that these programs would be beneficial to everyone, but the main focus was on the first-line supervisors. They specifically put the word "job" into the title of each program to show that they are practical and could be used by anyone. They made the formats simple (each has a four-step method), which made them easy to follow, easy to learn, and easy to use.

Today, we are not in the depths of an emergency that occurred during World War II and so we have more resources. As a result, we can involve all personnel in these programs. A major difference between the TWI programs and all other improvement programs is that, because of their design, anyone can quickly learn and use the concepts and method of any of the J programs. There is a plethora of courses and books on Lean, Six Sigma, Constraint Theory, and so forth. As much as the people who drive these programs want to get everyone involved, it is difficult because there is much to learn before one can use the concepts. The point is not that people cannot learn other programs. The point is that other programs take longer to learn. In addition, other programs deal with a higher level of production than that which an operator sees. A Value Stream Map, for example, looks at an entire product line whereas JIT and JMT usually deal only with the tasks within the realm of an operator. This is not to say that one is better than the other, just that they are different. Because they are different, other programs rely on a champion to drive the improvement. A program improvement specialist will lead a group of people to perform a project such as a Kaizen event. I refer to this as a "top-down" improvement since the incentive usually comes from upper management. Since TWI enables all personnel to know and use the programs, ideas and implementations start at the production workstation and thus I refer to TWI as a "bottom-up" improvement program. The improvements are smaller and individually less consequential, but collectively they will overshadow "larger" projects.

TWI's Main Objective Is to Solve a Problem and Not Just Deliver Training

The four men who drove the TWI Programs all had training backgrounds to some extent. Training was merely the vehicle that they used for transferring the knowledge they had so that companies would increase their productivity. Their objective was not to deliver training but rather to get people to use a specific method, which would solve a known problem. Each program has objectives[2] and when you have met those objectives you are successful. If you are having difficulty using one of the programs, "back up" and ask yourself what you are trying to accomplish. The methods given in the manuals outline a preferred method that should be universal. However, your circumstance may be different enough that something else applies.

We should not train for the sake of training, but rather to solve a problem or improve a situation. Although this can happen in any organization, it is especially prevalent in companies that have a training department. The objective of many professional trainers is to get people trained. This is measured by recording that participants have attended a session or sessions where they have been exposed to given material. Sometimes, the measure includes having the participants complete feedback forms telling whether or not they learned anything and if they thought the sessions were worthwhile. Often, participants have to sign a form saying that they attended the training. At the end of a given time period, say a year, the training manager can say with satisfaction that "X" number of employees were trained in such and such. However, when costs must be reduced, the training manager hastens to find a justification for that training. If she or he had used training as a problem-solving tool instead of as an end in itself, justification would not be a problem. Using training as a problem-solving tool means that a problem or opportunity for improvement must first be identified. Having done that, it must be discerned that training is actually a solution. As previously stated, the TWI Service believed that poor training or lack of training was responsible for at least 80% of the problems we incur. If, for example, a given operation is resulting in more scrap than is expected, based on similar operations, the operation is analyzed. The result may be that the setups vary. When we determine that the cause is operator-related process issues, we apply JIT by breaking down the job and training everyone who sets up the process. We then measure the process to determine if the scrap rate has been reduced. If it has not, we return to our analysis, which may include going back and verifying that we have gone through the steps asking

1. What problem am I trying to solve?
2. Am I using the proper J program for the problem?
3. Am I properly following the objectives?
4. Am I properly using the method?

If it has reduced scrap, we record the cost savings. This then would be our justification for the training. At the end of a given period, the training and production managers should be able to point to cost savings related to training. If they cannot, they are not using JIT properly. The same thinking applies to JMT and JRT.

TWI Develops Confidence and Resourcefulness in How to Proceed, Not Standardized Solutions and Rules[3]

The TWI Service recognized significant variability throughout the contractors, but they needed to create a program that would be suitable for everyone. They thus could not write a program that would specifically fit one individual's or one company's situation. A solution specifically designed for one situation would not fit all situations. They thus wrote each program as specifically as they could, based on the principles they knew. They left it up to the company to accommodate the training to their individual needs. They did not offer standardized solutions and rules because there are no standard problems. Each problem varies with the company and the situation within that company.

Use the Three J Programs Together, as Needed

All three J programs should be learned by all personnel and used whenever applicable. They should be thought of as problem-solving tools and not just as time spent in training. In learning to be a mechanical engineer, for example, a person learns mathematics, the strength of materials, stress analysis, materials science, and chemistry. When a problem or situation is encountered, the engineer will use whatever subject or subjects are appropriate to solve it. She must learn all disciplines because she does not know when any one will be needed. Therefore, we must train everyone in all three J programs because we do not know when they will need to use any one of them. Actually, they will probably have a need for each one every day. For example, do not train everyone in the facility how to clear a jam on the copy machine. Train only those who have a need for that skill. Training others who will never clear a jam is a waste of time. If, however, people waste time because they have to wait for the one or two people who know how to clear the copying machine jam, that is a problem. When we analyze the problem, we find that we may need JIT to train additional people, JMT to determine if the copier needs repair, or JRT to resolve issues created by the break in service.

TWI Is the Foundation of Lean, but Does Not Encompass All of Lean

The J programs were designed for specific purposes. Because they are so fundamental and because they satisfy basic human needs, they also can result in other benefits that people do not usually anticipate. However, they do not include all Lean concepts. It has been said that even if the J programs are used properly, they will not make an organization "Lean." However, if the organization does not use the J programs, they will never become truly Lean. The programs create a firm base on which Lean concepts can be established. Recognize that they will not cure all of an organization's ills. Furthermore, work and discipline are required to use them correctly. It is much like any other skill you learn in that if you do not master the basics, you will never be a master. Here are some Lean concepts that you will *not* find in the TWI manuals.

- Value Stream Mapping: Value Stream Mapping (VSM) is done at a higher level than that at which the J programs are used. VSM identifies waste in a process and notes with "Kaizen bursts" what should be eliminated. It doesn't, however, tell you HOW to eliminate that waste, which is done with JIT or JMT.
- Takt time: Takt time is the production time required to meet customer daily demand and although the TWI Service was aware of time studies, they did not address the issue of how much time it takes to do a given job. It should be noted that since a Job Breakdown Sheet (JBS) describes a job in fairly fine detail, if the JBS is followed, most people would complete the task within a few percent of the Takt time.
- Flow: Flow is akin to eliminating turbulence in a production stream. In hydraulics, a greater amount of liquid can pass if the flow is smooth or laminar than if it is turbulent. The same is true in production flow. The concept of production flow, as known today, was not used in the 1940s although there were assembly lines for automobiles, locomotives, and B-25 bombers.

The Methods Should Be Applied according to the Stage of Development of the Individuals and the Organization

The general method proposed by the TWI Service in 1945 still applies to all companies today. That method is to

1. Sell TWI to executive management and obtain its full support.
2. Sell TWI to the remaining management, obtain their support and involvement.
3. Structure a TWI team.
4. Identify a desired improvement, apply a TWI program, and obtain results.
5. Grow the program, under control, throughout the company.
6. Repeat with a second J program and then the third.

This implementation method should be used in all companies, but in a relative few, it can be expedited because the company's culture is ready for TWI. The TWI programs are more than just training programs; they are culture-changing programs. How the TWI programs are applied depends on where you lie on a spectrum or curve. The resulting culture will be one of "bottom up" or stewardship where management recognizes the value of the ideas and the work of all employees, and acts accordingly. In these relatively few companies of which I speak (at the far end of the curve), management already has respect for every employee and a true value of everyone's ideas and work. It recognizes that they have two main goals. The first is to make a product and a profit, and grow the organization, while the second is to develop the employees. What is missing are the mechanics to combine these two goals so that one is not omitted for the sake of the other. Once the management in this group experiences the TWI programs, they recognize it as a tool to span this functional divide.[4] This group needs only to learn the TWI programs and start using them.

In most companies (in the large center of the curve), management has become "enlightened" so they are willing to respect all employees, but they truly do not know how to do it. They know each person's work is important, but they have not developed the skill required to demonstrate it. This group consists of willing adapters but will require more time than the abovementioned group. Accepting TWI is not usually a problem because the basic and initial benefits of the programs are intuitively obvious. That is, most people recognize, for example, that in-plant training is poor and that JIT is a superior training method. Only after the programs have been in use for a while and have yielded results does this management know how to demonstrate respect for the individual, his work, and his ideas. After several months or a year, they look back and realize that they have, in fact, developed their employees while they were reaping the initial benefits of improvements in quality, cost, and so forth. This was done not by sending them to classes, but by increasing their involvement in the organization, which is now evolving into a learning organization.

The third group, which is at the far end of the curve or spectrum, will require much more time or they may never be won over. The number of "improvement programs" currently available is overwhelming and because it is difficult and time-consuming to select the ones that offer value, the TWI Programs are considered another "program of the month." Even if a program (usually JIT) can be established so that it yields results, this management may not get close enough to the operation to see the higher benefits. This prevents them from learning how to truly respect people and see the value in their work and ideas.

The TWI J Programs Are Skill Based

All three J programs deal with skills: transferring one's knowledge, making improvements, and dealing with people. A skill is the ability to perform a task. The more we practice that skill, the better we become at the task. The J programs were designed to be both practical and straightforward so that everyone could learn and use them. Although a person may quickly understand the basic four steps, masterful execution will require practice with proper, correct feedback. Furthermore, practice must be continual because, as with any skill, if we do not use it on a regular basis, we will lose proficiency. A well-known saying around musicians is, "If I miss a day of practice, I notice. If I miss two days of practice, my wife notices. If I miss three days of practice, my audience notices."

A Key to Teaching Is Asking WHY

A good student asks "why" because he is curious and wants to understand. A good teacher asks her students "why" because she wants her students to think through the problem and arrive at the answer by himself, if possible. For a teacher, there is a fine line between asking "why" too many times and not enough times. Asking "why" too many times can result in frustrating the student. One must know when the student has enough information to "put it all together" to arrive at a solution. On the other hand, not asking "why" enough results in giving the student the answer, preventing him or her from actually doing the thinking.

People Want to Be Productive and Be Involved in the Organization's Operation

Although it is not stated in any of the TWI manuals, an underlying premise is that people are not lazy and they want to be productive. This is a concept that all trainers must embrace because it is certain that sooner or later one will encounter a person who appears to be avoiding a job and the supervisor or trainer must find out why this is so. Psychological experiments have been conducted that show that people will rapidly lose interest in work they believe is non-essential or meaningless. When people see an objective, they become more enthused and willing to continue. In general, most people want to feel like they are part of "the team" and when they do, participation improves. Also, people do not make mistakes on purpose but mainly because they do not know the job as well as they should. This is a large reason why JIT is so helpful in increasing morale.

The TWI J Programs Help Satisfy People's Basic Human Needs

When the TWI Service started to help the defense contractors, they asked them what was needed to help them increase their productivity. They continued asking this question over the five years that the Service was in existence and most of the feedback could be summarized in the "five needs" they espoused. Today, it is known that the three skill needs of instruction, methods improvement, and relations mirror three basic needs of every individual. Edward L. Deci and Richard M. Ryan, two motivational psychologists, formulated what they refer to as the Self-Determination Theory (SDT)[5] in an attempt to explain why people do what they do; that is, what motivates people. Knowing that the three J programs address the needs of Self-Determination Theory (SDT) helps explain why these programs have been so successful and why most people readily accept them. Refer to Chapter 1 for additional information on SDT.

These, therefore, are the main concepts to keep in mind and follow when using the TWI J programs. Modifying other factors in the delivery will be acceptable and will not harm the outcome if these principles are followed. Keep in mind that there are also principles specific to each J program that must be followed.

Terminology is an important part of communication and when we discuss training, we often do not define our terms carefully enough, which can result in confusion. The word "training" can cover a wide range of topics and activities. A basic definition of training is to get something or someone to perform in a certain manner. In an attempt to avoid confusion, I will use the following definitions in this book.

- A participant: A person who attends one of the 10-hour J program sessions
- An instructor: A person who has attended 10 hours of the JIT session and has satisfactorily written a JBS, and instructed someone using it
- A coach: A person who has attended one of the J programs and has sufficient experience in using the particular program that she or he can assist other participants
- A trainer: A coach who has studied the J program manual and has satisfactorily delivered at least one 10-hour session under the observation of an Institute Conductor
- An Institute Conductor[6] is a trainer who has sufficient experience with both delivering the 10-hour session and implementing it into an organization, that she or he can develop coaches to become trainers

Finally, a person who becomes a trainer must expend a significant amount of effort, so I believe it is not fitting to say, "I trained him." My preference is to say I "developed" a person to be a trainer and acted more like a coach.

Notes

1. www.oxforddictionaries.com/us/definition/american_english/principle.
2. JIT: transfers knowledge to a person who can then skillfully perform a job; JMT: enables people to improve productivity by implementing their own ideas; JRT: enables people to develop and maintain strong, positive personal relationships.
3. Training within Industry Service. 1945. "The Training within Industry Report: 1940–1945." Washington, D.C.: War Manpower Commission Bureau of Training, p. 49. The original quote is "Confidence and resourcefulness in how to proceed, not standardized solutions and rules, are developed."
4. Dinero, Donald A. 2013. "Spanning the functional divide." *Performance Improvement Journal*, Vol. 52, No. 9.
5. Deci, E.L. and Ryan, R.M. www.selfdeterminationtheory.org/theory.

6. This is also referred to as a master trainer. I prefer "Institute Conductor" because this original term better describes the position. That is, this person is one who conducts a TWI Institute (the 10-hour session). Although an Institute Conductor may develop into a "master," it is neither necessary nor likely that a trainer will become a "master" at TWI.

Chapter 3

Understanding Job Instruction Training (JIT)

One must learn by doing the thing, for though you think you know it, there is no certainty until you try it.

Sophocles

Job Instruction is *the* way to get *a person* to *quickly remember* how to do *a job, correctly, safely, and conscientiously.*

"Why" is an extremely important word because it helps us satisfy our innate need for knowledge and understanding of everything around us. When given a suitable response, it also helps us verify what it is we are doing. There is no best time to ask the question "Why?" and, indeed, we should be considering it continually to keep us on track. If we do not know why we are doing whatever we are doing, actions become meaningless and we become less involved, if not physically then mentally. Friedrich Nietzsche wrote, "If you know the why, you can live any how."[1] Furthermore, the question occurs at various levels in our learning. Considering Job Instruction Training (JIT), the first question we should ask is why we should use or even consider it. This is answered by considering the objectives and benefits of JIT and especially a major benefit, which is standard work. Once we agree with the need for JIT, we address *What* it is, *How* we should use it, and *Why* we should use it that way.

Why Use JIT

Objectives of JIT

The central objective of JIT is to get a person to quickly learn how to do a job correctly, safely, and conscientiously. People learn in many different ways, but we are looking for the fastest, most effective way. JIT is the most efficient and effective way to teach processes to any capable and willing learner. Therefore, JIT is not necessary because people can learn a job without it. The downside, of course, is that it will take longer, the final learned process might not be what we want, there may be some poor quality product created and some equipment damaged, and the learner may hurt himself. Instruction consists of a wide spectrum, with JIT at one end and no external instruction (all self-training) at the other end. The improvements in quality, productivity, safety, and cost are maximized with JIT and minimized with no external instruction. All the instruction methods in between will result in varying amounts of quality, productivity, safety, and cost. Since people learn in many ways, the only requirement is that they receive feedback on what it is they do. Without feedback, there is no learning. If there is no one available to help us learn the job (external feedback), we can learn it by ourselves using our own internal feedback. When learning something by ourselves, there are two main ideas to consider. First, it will most likely take longer to learn the job than if someone were instructing us because we will have to solve each puzzle we confront. Second, because we are using our own knowledge, intelligence, and creativity to solve these mini puzzles along the way to the end, the method we end up with usually will not be identical to that which had been done before. If we are careful and conscientious, the method will accomplish the same objective, but the method and time used to do the job generally will be different. Most likely, the final product will also be somewhat different. In many cases it will be acceptable but often it will not be identical.

 The TWI Service was well aware of how people learn and they realized that although previous methods had been acceptable, they were no longer so. The Allies were in what was euphemistically referred to as "an emergency situation." Time was of the essence. The main objective was to produce as many goods as possible in as brief a time as possible, that is, increase productivity. Of course, a subset of productivity is quality because if many goods are produced and a large percent fail, we have not increased productivity. People who had no previous experience in industry must be brought up to production speed as quickly as each individual was capable. A farmer who had some familiarity with hand tools might be brought up to speed faster than a housewife with no such knowledge. Instructing someone in how to do a job also included instructing them to the point where they could perform the job as well as a "first class man."[2] Thus, the instruction should be as short as possible, but the results should reflect the quality and productivity required. It would be counterproductive to train people quickly and have them create poor quality products. The Service summed up their objectives for JIT in this sentence:

Job Instruction Training is the way to get a person to quickly remember how to do a job correctly, safely, and conscientiously.

Although we are not in the extreme "emergency situation" as the country was in 1940, we still have similar objectives in improving productivity and quality. As a result, we do not have to use JIT, but omitting it from our procedures can be detrimental.

Benefits of JIT

The main benefit proposed for JIT was shorter training times, but the TWI Service expected more than that. They believed that 80% or more mistakes happen because people have not been trained properly. Consequently, productivity increased not only because of shorter training times and improved quality, but also because fewer people were getting hurt on the job, there was less wasted material and rework. Today, we have experienced all that to be true, but in addition we have improvements in communication, teamwork, and morale.

Communication improves because people now have a standard method to discuss their jobs. A standard method has been determined and everyone follows that method. They do so not because they are told, but because they are shown that it is the fastest and easiest way to do the job that results in the required quality. When they have another idea about this job, they know they are welcome to discuss it with others. Similarly, teamwork improves because objectives and methods have been quantified and aligned. This is especially important across shift boundaries where it's difficult to get one shift to speak and consult with another. Partially as a result of improved teamwork, morale improves because everyone is more likely to feel part of "the team." In addition, JIT builds competence and thus confidence in people doing their jobs, which enables people to feel better about themselves.

Once we know the principles of JIT and of a Job Breakdown Sheet (JBS), we can use them any time we want to transfer our knowledge to others. Note that it is not required to create a JBS every time we want to transfer knowledge. Once we know the principles and the method becomes habitual, we communicate better when we are attempting to transfer knowledge.

Standard Work versus Standardized Work

Standard work is important because it enables stability and continual improvement in the workplace. Although that may sound like a contradiction of terms, it really is not. Standard work is characteristic of a given job being done the same way, every time it is done, by everyone who does it. Standard work need not include the "best" way to do the job, although if the JIT method is used correctly, it will result in a method that is collectively thought to be the best way we know how to do the job at this time. If everyone is doing a given job the same way every time it is being done, everyone knows how it is done and how it will be done when someone else is doing it. This is what creates stability. The point of standard work is to have everyone do a given job the same way each time it's done and then allow/encourage personnel to add their ideas to improve the method. This is what creates an environment of continual improvement. In addition, if everyone is doing a given job the same way every time it is being done, mistakes or variations in the method and/or output will be much easier to see. Thus, it will be much easier to see errors and any other variations.

There is often confusion between the concepts of "standard work" and "standardized work." Some of this confusion occurs because people use the terms interchangeably but then discuss them differently. I find the first step in reducing confusion in any conversation is to define the terms we will be using. Note that as TWI is a foundational element of Lean thinking, standard work is a foundational element of standardized work. Standard work has already been defined. Standardized work is standard work, but also includes the inventory required to do that job as well as the Takt[3] time required. In other words, when we are employing standard work, we can look at any given job and see that it is always being done the same way no matter who does it or when it is done. Standardized work, on the other hand, includes standard work, but also considers the Takt time it takes to do the job and requires that sufficient inventory is available to complete the job.

Standard work is work being performed (actions) and is NOT a method written on paper (documentation). Many people have procedures written and a small group of people (sometimes only one person) decides that is the best way to do the job, and thus this method is our "standard." When the job is attempted in production, it is found that the method is not possible or feasible and thus begin the back and forth changes. The only way to get people to perform standard work is to give them standard training. Standard training accomplishes two objectives. First, it shows that the method is valid and second, it is a reliable way to transfer the method to the operator. *Stabilization cannot be achieved without standardization, and the only way to get standardization is with standard training. The only standard training is JIT.*

Creating Standard Work

Some people have told me that they are not ready to use JIT yet because they want to create standard work first. That tells me that these people do not understand JIT because JIT creates standard work. There are two situations for creating standard work: the product has been designed and is ready to be put into production, or the product is already in production but the methods used to make it vary from person to person. If the product has not been put into production yet, engineers should develop methods for making it. Having done that, they should write a JBS for each of those methods. Once that has been done, the engineers should use the JIT method to instruct those in production who will be responsible for instructing production workers. This will accomplish two objectives. First, it will verify the process, showing that it actually can be done as the engineers planned it. In reality, the engineers will usually modify the methods to some extent in order to actually instruct production workers. The second objective achieved is that standard work will now be in place. The JBS can act as a standard work document or an additional document can be used if additional information is required for documentation.

If the product is already in production, but everyone is using a different method, creating standard work is a little different. In this case, the instructor creates a JBS by watching a skilled employee do the job. She or he then gets consensus by checking with all concerned and with the standard documents (if there are any). Since the product is already being made, once the quality and productivity have been verified, the standard documents must be made to agree with the actual process. Standard work will be performed because the operators will have been instructed in one method with JIT.

Principles of JIT

The following are the concepts that must be maintained if JIT is to be used successfully.

Learn by Doing: Everyone Must Do What It Is They Are Trying to Learn: Not Just Talk about It

To learn a task means to acquire the knowledge and the skill necessary to perform that task. In addition, in JIT we also require the learner to understand *why* he's doing what he's doing and to remember all that is required. Whenever we perform an action, we create a memory of that motion. When we repeat the same motion, that memory gets reinforced. When we are learning a task, we attempt to copy the motion we are shown or told, but everyone knows that our interpretation of what we see or hear affects what we do. That is why feedback is necessary for anything we learn. Without feedback, there is no learning. So when the coach says, "Swing level," we think we know what that means and attempt to copy the instruction. If the outcome is as desired, we reinforce that memory. If the outcome is not what we want, we alter the motion until we accomplish the desired results and then reinforce that motion. If the coach could put the execution of that motion directly into our brains, we would not have to "learn by doing;" we could learn by data input. The more we perform the task as intended, the less we have to think about it, which is why we want to make sure that we are performing it correctly.

Part of "learning by doing" has to do with verifying that we have actually learned the task. When someone tells us how to do a task, we estimate our ability to repeat the motions based on our knowledge and experience. For example, most people know how to ride a bicycle. There is an interesting video on YouTube created by an engineer by the name of Destin, who publishes "Smarter Every Day." The particular video to which I am referring is titled "The Backwards Brain Bicycle."[4] Destin was given a bicycle that had its handlebars modified so that the front wheel would turn in the opposite direction of the handlebars. When given the bicycle, he estimated that he could quickly learn to ride it since he had been riding bicycles for many years and knew what to do to ride this bicycle. The key is to turn the handlebars in the opposite direction of a standard bike. Because he has been riding a bicycle for over 20 years, his muscle memory was so strong that it took him eight months to successfully ride the bike smoothly. As he continued to ride, his motions became smoother. However, when he went back to ride a standard bicycle, he had a similar problem. This time, however, it took him only 20 minutes to master the standard bike. It took that long for his original muscle memory to come back into play. Note that his son, who is about six years old, and has not been riding a bicycle as long, took two weeks to master the "backwards bike." The son's muscle memory was not as strong and, being younger, his mind was more accommodating. This concept is shown whenever we see someone doing something with which we are not familiar and which looks "easy."

Sophocles (497–406 BCE) was a Greek philosopher and poet among his other endeavors. The TWI Service noted one quotation attributed to him:

> One must learn by doing the thing; for though you think you
> know it you have no certainty until you try.[5]

Sophocles said that over 2400 years ago based on his experience and knowledge. Yet today we are proving that concept with research. Of course learning varies with both the task and the individual, but the large majority of people are tactile learners. They must actually perform the task before either they or the instructor knows they can do it. The reason for this is based on the concept of Key Points. Specifically, the idea that all Key Points are not observable prevents people from successfully copying tasks.

This concept can be expanded to include the entire JIT program. When people see someone instructing using the JIT method, they often believe they can do it without practice. Furthermore, they believe they know all there is about JIT. This is not the case, and fully appreciating the value of JIT does take some time. Thus, everyone who has any responsibility for JIT must go through the 10-hour session as a participant for best implementation. Do not expect someone who has not participated in a 10-hour session to support the implementation (even though they may).

Resulting Action: Have all learners perform the task being taught. Do not assume that a person can do something merely because they can explain it to you. Have everyone in the organization participate in the 10-hour session.

Use JIT When Training JIT

The JIT method is the best way to transfer knowledge to another person. Since delivering JIT is transferring knowledge, we should use the JIT method when delivering the JIT method. Follow the four steps and adhere to the motto. The trainer should do everything s/he tells the participants to do. (Refer to the "One-to-One Instruction" section.)

Resulting Action: The trainer should write a JBS for each day's actions in the JIT sessions and be aware that she or he is using JIT.

One-to-One Instruction: Not Group

The Job Instruction (JI) method requires that the instructor teach the learner on a one-to-one basis. If two or more people need to learn a task, it is more efficient and effective if the instructor teaches each individual separately. There are several reasons for this. First, everyone is an individual and has a different knowledge base, experiences, and mental performance than every other person. Thus, each person will learn at a different rate. Instructing several people at once slows the training for some while it overloads others. In addition, the instructor must relate to the learner to find out what they know and what questions they have. This is more difficult to do in a group since some people are more outspoken than others and some do not want to admit they do not know something. As already mentioned, the instructor should watch the person perform the task. Time will not be saved when the learners are demonstrating the job because each one must do it individually. The instructor can watch only one person at a time.[6]

Group participation does have its place in discussing a problem, exchanging ideas, and coming up with new questions. The final instruction, however, must be done one-to-one.

Note that the 10-hour session usually includes 10 participants. This is done for both economic and educational reasons. The first two days (four hours) is lecture and discussion. A group is satisfactory for lecture and is preferred for a discussion. This is the first part of the second step of the four-step method—Present the Operation. However, following this principle, we cannot expect the participants to fully learn how to create and use a JBS just by telling and showing them one. We must meet with each participant one-to-one and find out what questions they have, and whether they really understand the method. For the reason stated above, this cannot be done in a group setting.

Resulting Actions: The trainer must spend some individual time with each participant when delivering the 10-hour JIT program. Instructors must train workers individually. Changes to existing JBSs must be communicated to workers individually, as when instructing.

Presenting Information in Small "Chunks" and Whole–Part–Whole

The JIT method was designed to be given to 10 participants for 2 hours a day for 5 days. The main reason stated was that a 2-hour "meeting" would not interfere substantially with production. In addition, the material was kept as simple as possible because professional trainers would not always be available.[7] A key statement in the JIT delivery is that "Job Instruction is the way to get a person to quickly remember how to do a job correctly, safely, and conscientiously." The objective therefore was not only to get the person to learn the task but to learn it as quickly as that individual was capable. Underlying all of this is the concept that people do not learn a large amount of information all at one time, but rather in smaller pieces or "chunks." "Chunking" is a term used in educational circles to describe a facet of how people learn.[8] If you think of how you would memorize a poem, for example, you would memorize one or two lines at a time. Once you had those two lines committed to memory, you would proceed to the next two while repeating the first two you learned. The point is that we can absorb only so much information at a time. As a result, no matter what parameters you are faced with, you must be aware of the fact that you can give out information only to the extent that the participants can absorb it. That follows if you are dealing with only one participant or 10.

The 5-day, 2 hour per day format is the standard and will work for the general population. However, it is recognized that we often face constraints where this format is not possible. Recognize that although each participant spends only 10 hours in the training room, additional time is spent creating the JBS, preparing for the demonstration, and thinking about what occurred in the session. If the 10 hours are done in one day, there will be no time for this, and thus the training's effectiveness drops significantly. If 3 days are used, with the first day comprising the first 4 hours of "lecture" (standard days 1 and 2), it is possible for the demonstrations to be done on days 2 and 3. However, when the total hours are added up, there may be significantly fewer hours used with the 5-day format. In any case, the trainer must verify that each participant actually knows the material. (Refer to "If the Person Hasn't Learned …" section.)

Perhaps an even better reason for using the 5-day, 2-hours-per-day format is based on the saying, "sleep on it." Although it is not known exactly how sleep helps learning and memory; it is known that "sleep has a key role in promoting learning-dependent synapse formation and maintenance on selected dendritic branches, which contribute to memory storage."[9] Thus, when we sleep, our minds are still working by reviewing what we did during our waking hours. After the participants receive the first 2 hours on Monday, they may believe they won't think about the material until Tuesday, but the opposite is usually true even if they do not think about it when they are awake. Reducing the opportunities for sleep between sessions reduces the learning potential. A comment I often hear on the fourth or fifth day is, "Now I understand what you said on Monday!"

A corollary to the concept of "chunking" is "whole–part–whole." One might think of it as, "chunking" is what we want the instructor to do and "whole–part–whole" is how we want her to do it. "Chunking" is based on the fact that people can absorb only so much information at one time. If we are to give an individual many small ideas or actions that she can absorb, then we are basing the instruction on the fact that the person knows how to put everything together properly to perform the task successfully. Since that will not guarantee success in instruction, we first demonstrate the whole job so the learner can see the scope of what she is required to learn. Then we break it down into parts so she can absorb it. Once each part has been explained, we "reassemble" the actions into the total task. This principle is used on many levels and in many applications. One example is the book you are reading now.

The recommended number of participants for JIT is 10. If there are fewer (say 6), there will not be enough demonstrations for everyone to grasp the concepts. In this case, it is recommended that everyone do two demonstrations. If there are more than 10 (say 12 or more) the sessions may well go over the 2-hour limit. Also, 12 demonstrations can be overbearing to some people and might discourage them from using the method.

Small companies that have trained everyone in JIT will occasionally hire additional people. These people should not wait for a JIT session until a contingent of 10 employees is available. The trainer can work with only one employee if that is required. In that case it would be advisable to have the participant do two or more demonstrations since s/he will not see any others. The training can also be combined with actual instruction.

Resulting Action: The trainer and the instructor must gauge how much a person can absorb at one time and deliver only that amount of information. The trainer should follow the manual and maintain the 5-day format. For the instructor, large tasks can be broken down into multiple JBSs. If that is not possible, the learner must be given only several steps at one time, leaving an experienced worker to complete the job.

Correct the Learner as Soon as He Makes a Mistake

Scientists are still studying the mechanics of how we learn, but we do know that repetition plays a role. Hence the saying, "Practice makes perfect." We know we must repeat actions if we want to become better at performing them. It has been noted that if we repeat incorrect actions, those are the ones at which we will become proficient. Hence the revised saying "Perfect practice makes perfect."[10] When someone does something incorrectly, the more they do it, the longer it will take for them to correct that action. Hence it is important to watch a learner the first time they perform a task and correct any mistakes as soon as possible.

A consequence of this is that a participant in a 10-hour JIT session should not create his first JBS by himself. It should be done under the guidance of the trainer or a coach. (Refer to Chapter 7—Implementing TWI–JIT.)

When correcting the learner, the trainer or coach should ask questions, as opposed to giving correction actions. The trainer/coach should give the learner the opportunity to think through the step so the learner develops a thought process. Giving the correct action immediately prevents the learner from thinking by himself.

Resulting Action: The trainer/coach must monitor the learner until the trainer/coach believes the learner knows the correct actions to perform the task. The number of times the task must be repeated in front of the trainer/coach will depend on the task, the learner, and the trainer/coach.

Repetition

As mentioned, repetition is important. The learner should never perform the task just one time for the instructor. The absolute minimum number of times a task should be done is two, but that would be only if the task is extremely simple. The reason for this is that the instructor really does not know what the learner is absorbing at any one time. Usually, the learner will gain some additional information with every repetition. The standard approach is to have the instructor do the job three times and the learner do the job four times. However, the important criterion is to have the learner perform the task until the instructor knows the learner knows it.

When delivering JIT, the instructor has many opportunities to repeat concepts.

A corollary to repetition is that the instructor/coach must test the learner to determine that the learner knows the task and can perform it. This is discussed more fully later in "How to Deliver a JBS."

Resulting Action: The instructor/coach must not assume that the learner knows a task or concept when it is given only once or performed only once.

If the Person Hasn't Learned, the Instructor Hasn't Taught

The purpose of this motto is to put responsibility for learning on the instructor. It is easy to deliver poor training and when it is found the person cannot perform the task properly, say, "I trained him, but Joe is difficult to train." This motto should not stand by itself because there are extenuating circumstances that apply.

It has been said that jobs do not get done properly because the person can't, won't, or doesn't know how to do the job. When a person can't do something, s/he is physically unable to perform the actions in question. For example, if a job requires discerning colors and a person is color blind, the job will be impossible for the person to do. Telling the difference between an orange stripe and a red stripe on a resistor may seem easy for a person who can sense colors, but it is impossible for a person who is color blind. "Can't" therefore has to do with one's aptitude.

"Won't" has to do with attitude. Most people know they have to be trained to do a job, but periodically some people think that they know the task already or perhaps they do not want to do that job. Sometimes a person has recently experienced a trauma and cannot concentrate on training. A common comment from instructors is that a person does not want to be trained. It is my belief that everyone wants to know more information, but the reason many people do not want to be trained is because they immediately think of all the poor training they have received in the past. They do not want to expose themselves to more of that. Poor training is like a double-edged sword. The experience itself can be mentally painful for a variety of reasons. The instructor goes too fast, doesn't repeat, uses terms unknown to the learner, gives irrelevant information, puts information out of a logical order, and so on. The other edge of the sword is that the learner is expected to perform the task after the instruction. Even with poor instruction, many people can perform the task, but often mistakes will be made. Since the learner has been "trained," the mistakes are the fault of the learner. The candidate thinks, "Why should I put myself into that position?"

"Doesn't know how" has to do with prerequisite knowledge. Even the simplest job can require that the learner possess some skills. Learning how to form a corrugated cardboard box may require the use of a tape gun. Although a tape gun is a fairly simple tool, a person unfamiliar with one may have trouble. Another example is in mathematics. In order to learn calculus, one must know trigonometry. In order to know trigonometry, one must know algebra. In order to know algebra, one must know arithmetic; and in order to know arithmetic, one must know numbers. If one attempts to teach trigonometry when the learner does not know algebra, both the learner and the instructor will become frustrated. The prerequisite knowledge of algebra is required for true learning to take place.

Resulting Action: The instructor must check for the learner's aptitude, attitude, and prerequisite knowledge before instruction begins. When they are all present, the instructor is responsible for getting the learner trained. If the learner is not capable, does not have the desire, or lacks some required knowledge, the instructor has the responsibility to recognize that something must be changed before the instruction can be successful.

Break Down the Job: Use the Concepts of Advancing Steps and Key Points to Analyze, Understand, and Use the JIT Method

The concepts of Advancing Steps and Key Points are central to the successful use of the JIT method and thus should not be disregarded. Writing and delivering the actions of a job in a continuous stream will, in some cases, serve as instruction, but the learner will not have the true understanding of the job required when confronted with variations. The creation and use of Advancing Steps and Key Points will be discussed in the coming pages.

The "What," "How," and "Why" of JIT

Preparation for Using the JIT Method

Preparation is important in anything we do. Abe Lincoln said, "If I had six hours to cut down a tree, I would spend the first four sharpening my ax."

That has been restated by athletic coaches who say, "Failing to plan is planning to fail." But what preparation is needed when we instruct someone?

The TWI Service listed four "Get Ready" points on its JIT pocket card that briefly describe what should be done before the actual instruction begins. In reality, much more has to be done than just these four points, but since the main subject is the JBS, I will cover them here for the sake of completeness. The remaining activities will be covered under Implementation in Chapter 7.

The four points listed are

- Create a timetable
- Break down the job
- Prepare everything
- Properly arrange the workplace

Training Timetable

A training timetable is also known as a training matrix, skills matrix, or job matrix. Refer to Figure 3.1, which lists the personnel in a given area and

Bill Smith pump rebuild (today's date)	Breakdown number	Disassemble	Clean	Inspect/sort	Sub-assm. 1	Sub-assm. 2	Final assm.	Run/test	Personnel scheduling notes
White		x	x	x	x	x	x	x	
Nolan		x	x	x	x		x	x	
Black		x	x	x	x	x	x	x	Retirement (m/d)
Jones		x	x				x		
Green		x	x				x		Needs more training in cleaning
Brown		x	x	x			x		
Riley		x	x						
Production scheduling notes				Increased cap. By (m/d)					

Figure 3.1 Training matrix.

matches them to the tasks that are done in that area. The underlying concept is that training should be scheduled as all other important functions are scheduled. Inherent in that is the idea that instruction should be done for a purpose and not just because there is available time or money. The matrix graphically shows who is to be trained, for what job, and when that training should take place. As the example shows, even with just seven operators and seven jobs, it is too much for a person to keep track of unless that is their only responsibility.

This may have been a great revelation in 1940 when the TWI Service first introduced it, but today many companies make use of it. Searching for "training matrix" or "skills matrix" on the Internet will produce a plethora of companies offering everything from free templates to complete consulting services and courses. Today's matrices contain more information than the TWI Service version and, of course, people modify them for their own organizational needs. They are an extremely powerful tool used for running a department, but they create other benefits when the matrix is posted for everyone to see. As part of a visual management system, they enable teamwork by sharing valuable information. Refer to Appendix I for information on a contemporary use of the training matrix.

Break Down the Job

This refers to creating a JBS, which will be discussed in detail next.

Prepare Everything

The instructor must have the correct equipment, materials, and supplies available for the instruction. Using a screwdriver to open a box when a box cutter is the correct tool does significant damage to the instruction. By doing this, the learner is taught that the instructor either does not know what tool to use or doesn't care. In either case, those are poor traits to pass on. It also does no good to give an excuse for not having the correct tools, materials, or supplies because that just teaches the learner to make excuses. "Do as I do and not as I say" also is *not* part of good instruction.

Properly Arrange the Workplace

This "Get Ready" point has evolved into the 5S Program. Stated succinctly, there should be a well-identified place for everything and unless it is being used, the item should be in its place. In addition, there should be no more tools or materials available than what is needed for the job. As with the training timetable, there currently is much literature on this subject.

The Job Breakdown Sheet[11]

It could be said that the JBS is the heart of the JIT program. Its misuse and/or lack of use may very well be a large reason the JIT program fell into disuse in the United States toward the end of the twentieth century. Today, some people create a JBS by putting their Standard Operating Procedures (SOPs) or Work Instructions (WIs) into three columns and labeling that document a JBS. When it becomes unwieldy and does not deliver any benefits, they incorrectly conclude that their job cannot be handled with a JBS. They may then resort to using just the JIT four-step method or they may drop JIT entirely. If a JBS becomes too large, unwieldy, or too cumbersome to use, it is incorrect and is not truly a JBS. Using a correct JBS[12] and using it correctly is the most effective way to teach someone how to do a job. Before we discuss how to create a JBS, we should first clarify the two most important concepts in it: the Advancing Steps and the Key Points.

It must be remembered that anytime you do anything, you are using a process. You may not be cognizant of the process, but you are using one nonetheless. Someone can create a JBS for that process and instruct someone how to do it using JIT. And that is the most effective way to transfer that knowledge and skill.

Details: Identification and Order

When we set out to instruct someone, we generally know the job in question. We know the job well since that is one reason we have been chosen to do the instruction. But perhaps we know the job too well. Sometimes we know it so well that we perform actions routinely and without being aware of what we're doing or the tricks we use. When we are not aware of every important action we take, we obviously will not tell the learner. The better we know the job, the more details will be relegated to our subconscious. We'll know them and can explain them when asked, but we are not conscious of them when we perform the process. Since we are not conscious of these details, we do not transfer them to the learner. How, then, does the learner know them? They resort to "trial and error," which is the cause of much scrap, rework, and injuries. Therefore, we must identify these details and write them down so that we tell the learner about them at the appropriate time.

For example, I was watching a man put a terminal on an ethernet cable. After he removed the insulation, he smoothed the wires with his fingers. I asked him what he had done and he said that he had removed the insulation. I replied that he had done something after he removed the insulation, but he adamantly stated that he had not. I asked him to repeat the process so I could better see what he had done. Starting over, he cut off the uninsulated wires and then proceeded to remove the insulation as he had done before. Immediately after removing the insulation, without putting down the wire strippers, he smoothed the wires with two fingers. I stopped him immediately and pointed out what he was doing. He then explained why he smoothed the wire. He did this to prevent a stray wire from protruding beyond the terminal. If not noticed, a stray wire could cause a problem. Although he knew he was doing this, he was not conscious of it because it had been a habit for such a long time. If he were instructing someone in how to mount a terminal, he may have neglected to mention this action, yet he felt it was important enough to do it every time he connected a terminal.

Order

In addition to identifying all-important actions, we must also arrange them in a logical order in which we want them to be performed. Our mind is an interesting device. When we know how to do a given job, all of the actions required are stored there in what might appear to be a random fashion. As we do the job, our mind puts all the actions in the correct order so we do the job as we should. All the habits we require, like smoothing the wire, are inserted at the correct time. We don't think of smoothing the wire as we are cutting it or even as we remove the insulation. Our mind triggers that action just when we need it. So we don't mention the habitual actions, and we may forget to mention those that are not quite habitual when we should. We continue with the job to some extent and then remember to tell the learner the important action. Thus, we put actions out of order. That, of course, doesn't bother us too much, but it does give the learner something else to think about. "Since he mentioned it, it must be important, but I'm not supposed to do that here, I have to do it right after I …" This thinking conflicts with what the instructor is telling the learner at that moment. So, good preparation consists of making sure we have all the important actions and putting all of the actions into a logical order that will make it easier for the learner to understand and remember.

Advancing Steps and Key Points

The TWI Service wanted to simplify instruction, so they broke each job down into Important Steps and Key Points. Important Steps were defined as "a logical segment of the operation when something happens to advance the work." They defined a Key Point as anything in a step that might

- Make or break the job
- Injure the worker
- Make the work easier to do, that is, a "knack," "trick," special timing, or a bit of special information

When I was first shown a JBS and given these definitions, it seemed very reasonable. However, when I started writing specific JBSs, I began to have difficulty. I was reassured that I would at some point understand it and that I should be aware of the fact that writing a JBS is probably the most difficult part of the JIT method. The difficulty I ran into was that it seemed like some actions, which I was told were Key Points, advanced the work. They moved it forward. In the example above, I reasoned that smoothing down the wire was an Important Step because the expert always did it and if it were not done, there could be a difficulty. Thus, it must advance the work. As I delivered more JIT sessions, I found out that I am not the only person who has this problem.

One day I had an epiphany. If an Important Step advances the work when I do it, then the work must <u>not</u> advance when I do <u>not</u> do it. That means when you omit an Important Step, the job does not advance or it stops. This seems very reasonable and makes it much easier to select Important Steps.

But what about Key Points? How do I know if the action I have selected is actually a Key Point? Looking at the definition, a Key Point is an action that makes or breaks the job. That means a Key Point is something that affects the quality of the job. Also, a Key Point is something that injures the worker.[13] That has to do with safety. The third point of making the job easier to do has to do with productivity. A fourth category added since the original is cost. A cost Key Point is any action that affects material usage. That would include excess use of any materials or supplies and could also include excess wear of tooling and/or equipment. Thus, stating the definition of a Key Point more succinctly, it is "any action that affects quality, safety, productivity or cost." Furthermore, Key Points are NOT defined to advance the work. If we fail to do a Key Point, we cannot predict the result. The job may or may not stop, it may result in a catastrophe, or we may not even notice that it was omitted. Whenever the work ALWAYS stops when we do not do an action, that action is an Important Step. As we write the JBS, we want to include all the actions that we believe are important to have the job completed to our satisfaction. In addition, we want to eliminate any other actions in order to make it easier for the learner to become competent. If we limit ourselves to all actions that advance the work (Important Steps) and all actions that affect quality, safety, productivity, or cost (Key Points), we will give the learner just the information s/he needs but no more. That means that if an action or idea does not advance the work <u>and</u> does not affect[14] quality, safety, productivity, or cost, we will <u>not</u> include it in our JBS. *This Then Defines the Standard Work for Us.*

When writing a JBS, people often question whether an action should be included. The first question should be, "If the action is not done, will the job stop?" If the answer is yes, then the action is an Important Step. If the answer is no, then the next question is, "Does it affect Quality, Safety, Productivity or Cost?" If the answer is yes, then it is a Key Point. If the answer is no, then it is not included at all.

A common example of confusion with identifying Key Points is in electrical work. Most people recognize that it is important to disconnect electrical power before working on an electrical system. Thus, they usually say that turning off the power is the first Important Step. Although it is important that we do that, it's actually a Key Point. What's drummed into our heads is "safety is always a Key Point," but we really do not know why it is a Key Point. The reason it is a Key Point is that we can continue to work without turning off the electrical power. If we do not turn off the power, we may successfully do our electrical work or we may get shocked with electricity. The amount of electricity we receive may cause us to drop a tool or it may be enough to knock us off a ladder or kill us. We cannot predict the outcome if we do not turn off the power. Otherwise competent electricians have died because they did not turn off the power before working on an electric system.

Another common example is changing a tire on a car. When asked for steps, people often say that you must loosen the lug nuts before you raise the car. However, if that is not done, the job really doesn't stop; it just slows it down (productivity). If the car has been raised and you find you cannot loosen the lug nuts, you would lower the car and loosen them when the ground prevents the tire from turning. However, *removing* the lug nuts is a step because if that is not done, the tire cannot be changed—the job stops.

When identifying actions as Key Points, keep in mind that a Key Point should represent only one idea. For example, "finger tighten hardware with heads facing out" contains two ideas. The first is to tighten the hardware "finger tight" and the second is to make sure the heads are facing out. If we combine those into one Key Point, the person may remember only one action. For this step, he knows that there are two Key Points (actions) to remember.

Some people use icons signifying the type of Key Point written down. For example, a blue "Q" might represent a quality Key Point and a green cross could represent a safety Key Point. When asked why the icons are used, various reasons are given, but the gist is that they believe some Key Points are more important than others. I believe all Key Points should carry the same weight or influence. Either they should be done or they should be ignored. The counter argument is that omitting some Key Points may only create an annoyance, while omitting others may cause a catastrophe. If omitting a Key Point has such harmful repercussions, then the job should be mistake-proofed to eliminate or reduce the impact of that Key Point. If we start grading Key Points as to their severity, then the ones that cause the least harm will soon be forgotten and the quality, safety, productivity, or cost of the job will degenerate.

Why separate actions into categories?

The idea then is to consider all actions in a job and place them into one of three categories:

1. Actions that advance the work
2. Actions that affect quality, safety, productivity, and/or cost
3. Actions that do neither of the above and should be ignored

Some people ask WHY we take the time to categorize all the actions into these three main groups. Why do we not just list the actions we want the learner to do and have him follow the list? That would actually be simpler, but the reality is that it does not work. People remember the Important Steps because they complete the job. If they miss an Important Step, the job stops, and that is very noticeable. People quickly learn what is necessary to complete the job. Key Points, on the other hand, do not always stop the job when they are missed. We separate out Key Points so that people will know to remember them. We want them to know what actions affect quality, safety, productivity, and cost, while not advancing the work. If the learner forgets to do a Key Point and something bad happens, she or he will surely remember to do it the next time. For example, if the apprentice electrician forgets to turn off the power and he gets a jolt of electricity that burns his hand, he will most likely always remember to turn off the power before starting his work. However, if he receives no shock at all while working, he may forget to turn off the power the next time he works. This may become a habit until he experiences an accident. In addition, if he avoids an electrical accident for an extended period of time, he will be reluctant to change his behavior when told to do so. Why should he turn off the power? Nothing has happened to him so far. People will learn all (or most) of the Key Points of a job if they do the job long enough. A purpose of JIT is to teach the learner all the Key Points initially so the learner will produce the desired quality, productivity, safety, and cost objectives when they start the job and not years later.

A second reason for identifying Key Points is that they show us where the variability is in a job and thus where someone can make a mistake. The Important Steps are WHAT we want the person to do, while the Key Points are HOW we want the person to do it. If we are telling the person HOW we want something done, that implies that there is at least one more way to do that step. Since there are several ways to do the step, that means the step has some variability to it. Note that there may be several ways to do a step, but if none of them affects quality, safety, productivity, or cost, we will not mention them. Therefore, we can use the list of Key Points as a basis for continual improvement efforts. We can also use the Key Points as a basis for corrective action. When someone does make a mistake, we can quickly identify the action that was missed that led to the mistake. Keeping Key Points separate from the actions of Important Steps makes it easier for a person to identify and remember where mistakes can be made, and it makes it easier for everyone to identify and correct errors.

It is also important to identify actions that should be ignored, since one of the failings in instruction is delivering irrelevant information that has no value but either confuses the learner or makes it more difficult for the learner to remember.

Terminology

We now know better what to write down as we watch an expert perform a task. Any action that will stop the job if it is not done is an Important Step.

Any action that affects quality, safety, productivity, or cost is a Key Point. Any other action is ignored and not included in the JBS. With this knowledge, JBSs no longer would be mysterious and seemingly illogical.

As I went from client to client, I found people were more quickly picking up the concepts but not as fast as I thought they should. In breaking down jobs, people would still include Key Points as Important Steps. In order to have the action in the correct column, I would ask, "Does it advance the work?" If the response was "Yes," my follow-up question always was, "If you did not do this action, would the job stop?" After several repetitions of this, the second question would not be necessary and I could stop at "Does it advance the work?" A comment often made is, "But it's important!" And I agree that it is an important action but that only means we should not ignore it. What was causing the confusion was the use of the word "important." When I first learned of JIT, I questioned the use of "important" as an adjective for steps. I learned that Toyota uses "major" instead of "important" and I asked some of my TWI friends their opinions. The thinking was that changing "important" to "major" would not reduce confusion because "major" is merely a synonym for "important." Also, that might imply that Key Points are minor. That, of course, is not the case because everything included in a JBS is required and thus important. It occurred to me to call the actions that advance the work "Advancing Steps" because they are unique in that they advance the work. I asked some people who have used JIT for a while what they thought of changing "Important Steps" to "Advancing Steps" since that better describes what they do. The response was that they did not think it was a good idea because they were quite familiar with the term "Important Step." When I broached the question to people who were just learning JIT, they were more adamant about accepting it. If the step advances the work, let's call it an Advancing Step! This reduces the confusion when learning the JIT concepts. Everything we include in the JBS is important or we wouldn't include it. If we understand the concepts, it does not matter what we call something. However, wording used throughout an organization should be consistent to avoid confusion. I will leave it to the reader to decide on the term to use.

Creating a Job Breakdown Sheet[15]

There are some requirements when creating a JBS. First is that it must be done by watching a person do the job. We want to watch the "expert" because we want to observe all the Key Points used to perform the job successfully. We often refer to this person as an "expert" since we want to watch someone who knows how to do the job well. The ideal candidate would be someone with the best productivity (quality and rate) record. This is the person we want to emulate and have all others in the department copy. This will not be the only person who has input on the JBS, but when dealing with one person, we will save time by using the most productive person available. Note also that the ideal number of people creating a JBS is two: one doing the job and one observing and writing the JBS. The JBS will not be as good if there is only one person both doing and writing. Also, if there is more than one person observing, the process can take longer.

Watching the "expert" perform the task is important for manual operations, but it is even more critical when performing tasks involving interactions between two people.[16] Doctors and social workers must interview patients in order to obtain personal, sensitive information, such as sexual history. Human resource professionals must interview people to get information about situations about which the individual may not be eager to divulge. Let's consider a JBS for an interview like the kind mentioned. The Advancing Steps in such a task are WHAT the information is; the Key Points are HOW the information is obtained. The required information is usually readily obtained and can be verified by consensus. In fact, a checklist can be incorporated into a form, making sure that nothing is omitted. However, one person may use that form to obtain all the information while another leaves the interview with an incomplete form. The difference lies in the Key Points and HOW the questions are asked. In watching the successful interviewer, it may be noticed that she or he enters the room with a smile (or not), maintains a certain distance, sits or stands in a certain location, and so forth. We do not know which of these points are important, so we must ask about each one. For example, it is probably well known that most people are more willing to divulge their medical information to a person wearing a white lab coat (a stethoscope around the neck is a nice touch) than to a person who has worked up a full sweat and is wearing a tee shirt, shorts, and tennis shoes. The former could be a new intern, while the latter may be a renowned cardiac surgeon who was rushed in during his day off to assist with a diagnosis. Skilled interviewers know that there is a difference in response whether or not there is a desk between the interviewer and the interviewee. Lack of furniture between the two puts them on a more equal basis while speaking over a desk adds some power to the person whose desk it is. We do not want to make clones of everyone, but there are Key Points people use that assist them in successfully achieving a task. Those are the Key Points that we want to identify and copy. The term for this in the medical field is "bedside manner." As with obtaining many Key Points, this is not always an easy thing to do, but it does become "easier" with experience. Also, the same questions are asked of the "expert" in order to determine the Key Points. A conversation such as the following might take place.

JBS writer: Why did you sit down immediately after you entered the room?
Expert: I can get better answers faster because it's less intimidating to the patient.
JBS writer: What would you do if there were no chair on which to sit?
Expert: I know why I'm interviewing the patient, so for this task I check to see that there is a chair in the room before I enter. If there is no chair, I have someone else bring one in before I enter.
JBS writer: Would you sit on the bed if there were no chair?
Expert: No. That would be an invasion of the patient's space and inhibit conversation in many cases.

Every organization must determine for itself the best way to obtain these Key Points. Perhaps the "expert" could be videoed interviewing either an actual person or an actor. The video could then be reviewed multiple times to discover all the Key Points. Keep in mind that a JBS should be "a living document" and as Key Points are found, they should be added. Conversely, if a Key Point is considered to be not effective, that action should be eliminated.

The second requirement is that the expert must actually do the job at the job site. Since you are choosing an experienced person, she or he can undoubtedly tell you everything about the job. However, if you do not see the workplace, you will not be able to ask questions about materials, tools, or actions that the expert takes for granted. Do not get the job information for a JBS from a WI or SOP. You want to see how the job is actually done so that you can identify all the Key Points. Continuing with the interviewing example above, if a doctor were asked about the interview from his memory, the question of a chair may never arise. You should refer to the WIs or SOP after the JBS has been written to verify nothing has been omitted, but the first draft should be done while watching the expert because that is how the product is being made.[17] If the JBS does not match the WI, either the product is faulty or the WI is incorrect. Make sure the expert has sufficient time to work with you in writing the JBS, since you will be slowing down the operation asking him to repeat and do steps over.

If you have never seen the job before, ask the expert to perform it once so you can see the scope involved. When watching for the first time, you will start to formulate an idea about the steps and Key Points.

Video

Some people ask about using video in documenting a job in order to write a JBS. In reply, I refer to the idea that technology is good but it must be used for a defined purpose. There are two main reasons for using video when writing a JBS. The job should take only a few minutes so as not to overload the learner, but writing the JBS draft may take much longer. First, the job can be repeated as often as you like so that you can better know what is being done. This might be required if you want several people to view the job in getting consensus, or if the job must be performed many times to select all the Key Points. It would also allow the JBS to be finalized over a period of days without severely impacting the "expert." The "expert" must still be questioned, but his or her questioning time can be minimized. An example as noted above could be for an interview, where it might be difficult to discover subtle nuances through just one viewing. A second reason for using video is to slow down the operation or speed it up for clarity. High-speed photography, for example, is useful when a person's hands are moving so fast you cannot really tell what is happening. When this occurs, the expert often cannot explain what s/he is doing.

The disadvantages must also be considered. Although it might be thought that the operator does not have to be interrupted when taking a video, that is not completely true. The time it takes to write the JBS may be longer because you inevitably must return to ask some questions. You may even have to return to watch from a different angle or take an additional video. The best way to write a JBS is to watch the operator live. Generally, more can be seen watching a live operation than can be seen through the filter of a camera. There are exceptions, but one should use a video only when circumstances show it to help.

An overarching idea to keep in mind while writing a JBS is that the purpose is to create notes for the instructor to use during the instruction. We want to include all the necessary information the learner needs, but no other information. That means we do not want to include what the learner already knows. Since we do not know who the learners will be as we write the JBS, that seems like an impossible goal. However, with experience, we can estimate what the average person should know, keeping in mind that the

JBS can always be changed as needs arise. For example, in the tire changing JBS, there is no Advancing Step for opening the trunk. The thought here is that if the person does not know how to open the trunk, we may be instructing at a level above which the learner is capable and we should not continue with this JBS.

The JBS "Changing a Tire on an Automobile" in Figure 3.2 will be used as the example when discussing the various aspects of a JBS.

No.

JOB BREAKDOWN SHEET

Operation: Changing a car's tire
Parts: Replacement tire
Tools & materials: Tire iron (wrench & flat blade); jack; rubber hammer; manual; car keys; tire pressure gauge
Common key points: (4) Park on level surface; put car in 'park' or in gear; engage parking brake; chock at least one wheel; = stabilize car
Technical terms: Hub = part of the car that holds the wheel and tire; Wheel = metal part that mounts to the hub and holds the tire; tire = rubber ring that mounts on the wheel and supports the car on the road

#	ADVANCING STEPS	KEY POINTS	REASONS
	A logical segment of the operation when something happens to advance the work	Anything in a step that might— 1. Make or break the job 2. Injure the worker 3. Make the work easier to do, i.e. "knack," "trick," special timing, bit of special information	Reasons for the key points
1	Raise car	1. Air in spare 2. Remove hub cap w/blade 3. Loosen nuts 4. Raise enough	1. If low, decide when to fill 2. Proper tool 3. Weight of car provides resistance 4. A flat tire smaller profile than a full tire
2	Remove nuts	1. Place in hub cap	1. Don't lose
3	Remove tire		
4	Replace tire	1. Clear any debris 2. Line up holes on ground 3. Lead with top 4. Wheel flush with hub 5. Tighten with tire iron 6. Alternate studs	1. Debris may fall out later and tire may loosen 2. Takes less time & effort 3. Use gravity to hold tire 4. For proper tightness 5. Wheel doesn't shift when set on ground 6. Prevent warping of wheel
5	Lower car	1. Check manual for torque 2. Tighten nuts 3. Replace hub cap with rubber hammer	1. Nuts won't loosen 2. Per torque 3. Prevent damage
	WHAT	HOW	WHY

Figure 3.2 JBS for changing a car tire.

The Heading

The heading is important for several reasons. The name of the job is important because it tells the learner what he is going to be doing in the job and how to think about the job. An example of this is when we were writing a JBS for a position where a machine would cut tubing to length, and an operator would remove the part and place it into a container for the next operation. The job title was Machine Unload and it had just one Advancing Step: Unload tubing. It did have several Key Points:

1. Check for burrs on ends
2. Check inside for contamination
3. Check length every 50 pieces

Although there were only three Key Points, operators would occasionally miss looking for the contamination. If there was contamination, they were to remove it with a brush. If it happened too often, they were to notify their supervisor because something was wrong with the process. We changed the title of the job to Tubing Inspection. The Key Points were now Advancing Steps because if the steps were not done, the inspection job would have stopped; it would not be complete. More importantly, however, is that the operator now thought of the job as an inspection job. He was now an inspector and not an unloader and it was easier for him to remember the three Advancing Steps. Consequently, the title of the job should really tell the learner what you want to be done.

That can cause problems in some cases because the operators may already refer to the job by some name that may not exactly describe the operation. For example, in some assembly line operations, the line must be cleared completely of one order before the next order can start, in order to prevent mixed batches. This is especially true in industries regulated by the FDA and other agencies. All parts must be traceable and their locations known at all times. All inventory must be accounted for and all orders must contain the exact number of parts. In order to do this each order must

- Be verified physically to have the correct number of parts
- Have all accompanying paperwork (routers, work orders, etc.) be in agreement
- Have the assembly area physically cleared of all parts
- Have all overages returned to stock
- Have all discrepant material accounted for and processed

These are all individual jobs and each would have its own name. However, if one person does the first two and another does the last three, either person might refer to what they are doing as "line clearance." In naming a JBS for jobs one and two, we might call it "Order part count and router check," but that may not be quickly accepted by the person who has known the job for a while as "line clearance."

In the heading, we also include all the parts, tools, equipment, and materials required, all references, and any other pertinent information that will help us set up for the job. This acts as a checklist so we know exactly what we need. In addition, we can add any fields to the heading that will help us with the instruction. For example, we might include some background notes that we want all instructors to share while they are explaining the job in Step 1—Prepare the Worker. We might include a field for terminology that we believe most learners will not know. Depending on what filing system we use for our JBSs, we will probably include a number or alpha-numeric identification so that we can easily file, retrieve, and modify the JBS as needed. Each organization must determine for itself the system that works best. Factors to consider in creating an identification notation would be, for example, the type of technology to be used (loose leaf notebooks, filing cabinets, computer systems), the number of facilities (single site versus world wide locations), and who has access and for what purpose (everyone for everything or levels for read-only, instruct only, approve only, modify, finalize, etc.)

Refer to the JBS of changing a car tire (Figure 3.2) to see an example. Note that there are many differences among automobiles and this JBS does not account for all of them. When training someone in a job that has as much variation as this, the training would start on a given design such as one similar to that the person drives. If the intent is to train someone as a mechanic, once the person learned this JBS, modifications could be added.

The only part needed is the spare tire and the tools are those commonly found in the automobile. If the learner were being trained to be a mechanic in a garage, the instructor might start out with the manual wrench supplied in the car and then progress to a pneumatic wrench found in a mechanic's shop.

Four Common Key Points are listed. These are Key Points that apply to each Advancing Step. Instead of repeating these four Key Points in every step, we call them "Common Key Points" and tell the learner that they apply to the entire job and must be done with each step. In this case, each of the Common Key Points has the same reason—to stabilize the car during the job.

Technical terms are also listed here in the event the learner is not familiar with them.

The Body

The main body of a JBS consists of two columns of Advancing Steps and Key Points. Another way of thinking about the actions in a JBS is that an Advancing Step is WHAT we want the learner to do and a Key Point is HOW we want him to do it. If we are going to tell the learner HOW we want the step done, we should also give him a Reason WHY we want it done that way. Hence we now have three columns of information.

Note that the inclusion of reasons serves two purposes. First, it helps the learner remember the specific Key Point by providing the reason(s) we do it this way. If I am given a Key Point, "wear safety glasses," and the corresponding reason is "particles in the air could lead to loss of eyesight," I will have an understanding of why I should wear the safety glasses and also much more of an internalized interest in using them. When we know the consequences, we are more likely to follow the procedure. The second purpose of a reason is that it validates the Key Point. If there is no reason for a Key Point, there should be no Key Point. If the expert says, "That's how I was trained" or "That's the way we've always done it," you must dig deeper and find out why. If there is no reason, there should be no Key Point. This is not to be taken lightly. Just because the expert does not know the reason for what he thinks is a Key Point does not mean that it does not exist. A valid reason must be found. If there is no valid reason for a Key Point, there is no Key Point. We should not be doing something because we always have done it. In some cases, it may not be easy to find a reason, so the "Key Point" should be done until a decision has been made. Often, procedures or designs change and some processes are not brought up to date.

Recording Advancing Steps

The first objective is to find all the actions that can be classified as Advancing Steps. There are three main ways of looking at a task when we are analyzing it in order to create a JBS.

1. Manual
2. Visual
3. Software

The manual category is by far the largest and the one people think of when breaking down a job for JIT. I use the term "manual" because we are physically doing something such as assembling, completing a form, or doing any of the productive jobs we do every day. The visual category has to do with inspecting or using a checklist, for example. When we have finished the job, nothing may look different; but if we have not completed each step, the job has not been done correctly. The software category is thought of when we are using a computer and the software in it. I am using three distinct categories only in order to explain the concept of finding Advancing Steps. There are some tasks that will fit into each category very well, but there are others that might have characteristics of more than one category. You might have to check the criteria for each category before you are satisfied that you have found a reasonable Advancing Step. I will discuss the first two categories here and explain the software category under its own heading.

When Key Points arise, they can be noted if you think you might forget them, but focus on getting Advancing Steps first. This will give you a framework for the JBS. The way you do this is to have the person start to do the job and have him stop when you think you have come to an Advancing Step. You have seen the job at least once so you have an idea of what the steps might be. If you're not sure, have the person stop anyway. For a *manual job*, the questions to ask are

- "What did you do?"
    - We ask this to make sure we are both talking about the same action.
- "Would the job stop if you did not do this?"
    - We ask this to determine if the action is an Advancing Step.

When you ask these questions, you should already have an opinion. For example, in the job of changing the tire on a car, the expert is standing next to the car and you would ask him to start the job. He walks to the rear of the car and opens the truck. You might ask, "What did you do?" The exchange would be

Expert: "I opened the trunk."
You: "Why?"
Expert: "I have to get the spare tire and the tools."
You: "Would the job stop if you did not do that?"
Expert: "Yes."

We want to keep the JBS as sparse as possible so that we tell the learner only what s/he does not already know. Although the job would stop if the operator did not obtain the spare tire and the tools, is it necessary to tell the learner that? Do not make the decision yourself, but have a discussion with the expert to determine if this should be included as an Advancing Step. In this case it was decided that the learner would know where the tools are and/or it will be obvious enough when we do the demonstration. In addition, opening the trunk is fairly straightforward and needs no further explanation. However, if something does need an explanation, we would include it as an Advancing Step. You must use the experience and judgment of both yourself and the expert in making this determination. For example, most cars have their spare tires in the trunk, while minivans and pickup trucks often store the spare tire under the floor outside of the body. In this case we must decide if all learners would know how to access the spare tire just by watching or if additional words are required. If additional words are required, then we must add an Advancing Step (and later the required Key Points.) The concept is to make the best decision possible and then adjust the JBS as you gain experience in instructing people on the job. Much information in many jobs is obvious to the learner, so we should not tell him something he already knows. We should restrict our dialogue to what the person does not know. Most of the time, you and the expert will be correct in your decision.

You continue to watch the expert perform the job and arrive at the first Advancing Step of "Raise car." If this is not done, the tire cannot be changed and the job would stop. In addition, this is WHAT you want done, but there are many ways to do it, so you must tell the learner HOW you want the car to be raised.

We must think a little differently when we are doing a *visual job* such as inspecting or completing a checklist. In these cases, eliminating any step will not stop the job, but it will make the job incomplete. For example, we might have a JBS titled "Final Sub-assembly 1 Inspection" that has the following Advancing Steps:

1. Check label
2. Check surface finish
3. Check trabe operation

Any two of these steps could be done if the third one is not done, so they do not really stop the job when omitted. However, if any step is omitted, the job is not complete.

Collect all Advancing Steps until the job is finished and then review the steps with the expert. In this spare tire example, we would say,

We've got five steps that are

1. Raise car
2. Remove lug nuts
3. Remove tire
4. Replace tire
5. Lower car

Notice that, in writing the Advancing Steps, we should use as few words as possible and try to limit wording to a noun and a verb. Instead of "Use the jack found in the trunk to raise the car," it has been simplified to "Raise car." We eliminate articles such as "a" and "the" and any other non-value-added words. Advancing Steps should be thought of more as a title than a description of an operation. We do this because we want the Advancing Steps to give an outline of the job that the learner can easily understand and remember. We have told the learner WHAT we want him to do. Now we have to tell him HOW to do it.

Recording Key Points

Finding Key Points is a skill or an activity that improves with practice. Some Key Points are obvious to you, the expert, or both of you. Other Key Points are subtle or happen so quickly that they are easily missed. There are some general guidelines that you can use to find Key Points.

By definition, a Key Point is any action that affects quality, safety, productivity, or cost. The procedure then is to ask the expert to do the job one step at a time and determine if either of you can associate Key Points with that step. In many cases the expert will know most of the Key Points, but sometimes she will perform a Key Point so habitually that she does not recognize she is doing it. For example, when the expert is doing Step 2, "Remove lug nuts," she may put them into the hubcap without thinking. You might then ask, "I noticed you put the lug nuts into the hubcap after you removed them. Why did you do that?" If the answer is that it's just a habit and it does not affect quality, productivity, safety, or cost, then it should be ignored. If the answer is that they are easy to lose (affecting quality and safety), especially when you are on the shoulder of a road, then it is a Key Point. Other questions one might ask are, "What would happen if you did this?' or "Would it be easier if you did this?" or "Why not do this?" Note that we are not looking for improvements here, although they are often found. The objective is to identify all actions that affect quality, productivity, safety, or cost and ignore all actions that do not.[18]

When a reason is given, it can be written in the Reasons column. The expert must give a valid reason for every Key Point recorded, as described.

The order of Key Points is not critical but does facilitate instruction to some degree. Generally, Key Points should be in the order of the operation because it makes them easier to remember. Notice also that Step 1 "Raise car" has nine Key Points (four Common Key Points and five separate Key Points) and only the last Key Point is actually done while the Advancing Step is done. The other eight Key Points are done before the car is even raised. The JBS is not intended to record a procedure; its purpose is to assist the instructor in transferring knowledge. The Advancing Steps must be in order with respect to themselves and so should the Key Points. However, the first Advancing Step need not be performed before its first Key Point because of the way we have defined these terms.

The classic example to explain this is the job of changing a wall-mounted light fixture. The first Advancing Step is to remove the light's bezel because if that is not done, the fixture cannot be replaced; the job stops. The first action, however, is to turn off the power and the second action is to lock out the switch box. Neither of those actions must be done to change the fixture, but we are gambling with high stakes if we do not do them first.

Step 3 "Remove tire," does not have any special Key Points.[19] We are basing this on our knowledge and experience in seeing people change tires. Whenever anyone gets to the point of having the lug nuts removed, they always remove the tire the same way and we do not have to tell them how to do it. This has wide ranging implications. Since an Advancing Step is WHAT we want the learner to do and the Key Points are HOW we want him to do it, if there are no Key Points the implication is that we do not care HOW the step is performed. There usually are Key Points, but in some cases no actions apply that affect quality, productivity, safety, or cost. For example, Step 4 "Replace tire," has six Key Points. If none of those Key Points are done, we cannot predict the outcome. There is a good chance we may not see any effect. In Step 3, however, if the learner does the step, she will do it correctly. That means she cannot make a mistake if she does Step 3. What we surmise from this is that Key Points represent the variation in a job. They represent any place a person can make a mistake. They must have completed the job and thus done all the Advancing Steps. But if they made a mistake, they omitted a Key Point and we should include that action in the JBS. When ferreting out Key Points, we can now ask the expert where people have made mistakes on this job in the past. This may not only reveal additional Key Points, but it will also give us an indication what the average person will or will not do. For example, Advancing Step 5 (Lower car) does not include Key Points for removing the jack or replacing the jack into the trunk. The question to ask is, "Has anyone ever made a mistake by lowering the car and not removing the jack?" People who are learning to write JBSs often want to include every action because of their familiarity with work instructions. They may say that they do not know of that ever happening, but they know that it is possible. Many actions are possible, but based on our judgment and experience we want to include only those actions that the learner would not already know.[20]

Although most jobs done by humans have Key Points, it is possible to design a process that has only Advancing Steps and no Key Points. If this is the case, then this job would always be done correctly if it is completed. The mistake-proofing has been taken to the point where a mistake is not possible. If a person completes all the Advancing Steps, the job will be correct all the time. This is the objective for which we aim. If this is technically possible, barriers to achieve it would include cost effectiveness and time.

Changing Key Points

Once you start writing JBSs according to these definitions, you may find that the amount of Key Points can quickly increase. You may find that there are only one or two Advancing Steps to a job, but there may be 10 or 15, or more, Key Points. There has been much discussion about how many Key Points there should be in a job and the answer is "as many as the job dictates." The JBS is based on the process being used in the job, and that will define the number of Advancing Steps and Key Points. Some people have said that there should be a certain ratio of Advancing Steps to Key Points or only so many Key Points. The fact is that the process determines the number of Key Points, so if you want to remove any Key Points, you must change the process. Changing a process to prevent someone from making a mistake is called "mistake-proofing." Moving an action from the Key Point column to the Advancing Step column, without changing the process, will give you problems. Remember that if everyone does all the required actions in a job, then the job will be done correctly. The reason we are separating actions into two categories is to help people remember to do all of them. If they forget to do an action that is a Key Point, they may not realize it. Because the action is a

Key Point, we cannot predict when or if they will find out. The more time that goes by before they find out they missed an important action, the larger the chance will be that damage of some sort will occur.

An example of how writing a JBS can drive mistake-proofing is seen in the following example. Refer to the JBSs in Figure 3.3 for loading a bag. The situation is that a company that makes office chairs ships a model of chair with a bag of hardware attached with a wire tie to the column that supports the back of the chair. This JBS is to instruct operators how to fill the bag with the assembly instructions, the hardware, and a wrench, and attach the wire tie. This JBS is done in one part of the plant and then is taken to the shipping department where it is packaged. The original JBS calls for "attach wire tie" to be Important Step 3.[21] The first question would be, "Does that advance the job?" The person who wrote the JBS said that it is important and so should be included as an Important Step. He said that if the wire tie is not attached, the bag of hardware could leave the workstation and go to

No. 1237

JOB BREAKDOWN SHEET

Operation: Load hardware bag; Chair FXQ 1734
Parts: Wrench, instructions, #12563 screws, #96745 screws, instructions Doc 526
Tools & materials: zip bag #1564, hardware tool TR# 778; wire ties #774; tote
Technical terms:
Common key points:

ADVANCING STEPS	KEY POINTS	REASONS
A logical segment of the operation when something happens to advance the work	Anything in a step that might— 1. Make or break the job 2. Injure the worker 3. Make the work easier to do, i.e. "knack," "trick," special timing, bit of special information	Reasons for the key points
1. Load instructions	1. Fold in ¼ 2. Title shows	1. Fits into bag 2. Customer can see
2. Load hardware & wrench	1. Use tool – all holes 2. Fasten bag 3. Fasten tie	1. Confirm all pieces and amount are correct 2. Hardware doesn't fall out 3. Shipping can hang on chair
3. Place into tote		
WHAT	HOW	WHY

Figure 3.3 JBS for loading a hardware bag (first).

shipping. Once there, the packers would not be pleased because they do not have wire ties and would have to either return the bags or find some wire ties. I asked if bags were ever sent to shipping without wire ties. He replied that it happens enough to be a problem, which is why it should be an Important Step. I explained that because omitting the wire tie does not stop the job (it can still go to the shipping department), it is a Key Point. For this process, attaching a wire tie is a Key Point and one cannot arbitrarily say it is an Important Step. The fact that it is important is not relevant because every action we include in a JBS is important or we would not include it. In order to make the action "attach wire tie" an Important Step, we must change the process so that the job stops if the wire tie is not attached. I made the suggestion to change the process by changing the material handling device from a tote to a post with a peg. (Refer to Figure 3.4a.) With this as the material handling device, the operator would have to hang the bag on the peg in order for it to leave the workstation. If the wire tie were not attached, the bag could not be hung on the peg, the bag would not leave the workstation, and the job would stop. We thus changed "Attach wire tie" from a Key Point to an Important Step by changing the process (Figure 3.4b).

There are three learning points that come from this scenario. First, there is a difference between an Important Step and a Key Point. However, both are important or we would not include them. This gives us reason to call the steps Advancing Steps to avoid confusion with Key Points, which are also important. The second point is that when writing a JBS, we often see areas for improvement because we are scrutinizing the job. We do not go into the depth we would if we were using Job Methods Training (JMT), but a JIT analysis does reveal many opportunities for improvements. The third point is that a given process has an inherent amount of Advancing Steps and Key Points, which are defined by the process. We cannot merely move an action from one column to the other so we have a "good" number. The only way to change either an Advancing Step or a Key Point is to change the process and we usually do that through mistake-proofing.

Software[22] Applications

JIT was developed in 1940 when computers were unlike what they are today regarding capability, complexity, availability, and quantity. Many of the jobs we do today involve the use of a computer and a keyboard. Many devices today (video players, games, toys, appliances, and so forth) contain microchips to which we must give commands by pressing buttons and responding

Understanding Job Instruction Training (JIT) ■ 63

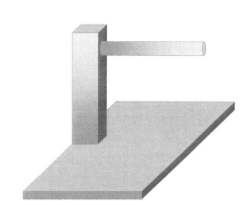

(a)

No. 1237

Job Breakdown Sheet

Operation: Load hardware bag; chair FXQ 1734
Parts: Wrench, instructions, #12563 screws, #96745 screws, instructions Doc 526
Tools & materials: zip bag #1564, hardware tool TR# 778; wire ties #774; ship pallet
Technical terms:
Common key points:

ADVANCING STEPS	KEY POINTS	REASONS
A logical segment of the operation when something happens to advance the work	Anything in a step that might— 1. Make or break the job 2. Injure the worker 3. Make the work easier to do, i.e. "knack," "trick," special timing, bit of special information	Reasons for the key points
1. Load instructions	1. Fold in ¼ 2. Title shows	1. Fits into bag 2. Customer can see
2. Load hardware & wrench	1. Use tool – all holes 2. Fasten bag 3. From and to locations	1. Confirm all pieces and amount are correct 2. Hardware doesn't fall out
3. Attach wire tie		
4. Hang on post		
WHAT	HOW	WHY

(b)

Figure 3.4 (a) Shipping pallet for hardware bag. (b) JBS for loading a hardware bag (second).

to lights. As a result, people ask if the JIT method can be used for computer applications or programming devices. JIT can be used whenever a person does something, so JIT can be used for jobs that use a computer or a microchip. However, we must look at how the method is applied. It is helpful to remember that a computer and a microchip are only tools intended to allow us to do a given job faster than if we did not use them. In order to address this application, the main focus should be in thinking of WHAT we are doing and then determine HOW we will do it with a computer. The JIT method is effective because in manual operations, people usually tell the learner WHAT should be done, but they often neglect to say HOW it should be done. As mentioned, people make mistakes because they do not perform Key Points and JIT corrects this by specifying all Key Points. In a software application, however, people usually tell the learner *How* the job should be done (which keys to press), but they neglect saying *What* they are doing. Knowing *What* to do in a software application really imparts understanding of the process and it makes the *How* easier to remember.

Let us first consider the heading. We must label the job, so the field for "Operation" is still necessary. The second field, "Parts," is not really applicable because we often do not have any when we are working with a computer. The main interface between the actual computer and the computer operator is the software the operator uses to control the computer. There are many competing software programs available that perform similar operations, but knowing how to use one does not guarantee the person knows how to use another similar software. For example, Microsoft Excel and MAC Numbers were both written to store, retrieve, and analyze data. However, knowing how to use one does not guarantee knowing how to use the other. The same is true for larger software systems such as Oracle and SAP. We should thus have a field describing what software system we will be using in order to remind the instructor to check the learner's knowledge about it. I use the term "System" to be all-inclusive. The JBS will be written based on a knowledge level of the learner, and the instructor must check for that. In that regard, some people like to add an additional field for "Prerequisite Knowledge" so that specific parts of a system can be noted. For example, the system may be "Microsoft Office" and the prerequisite knowledge might be "extensive use of Excel."

In addition, we are not assembling, modifying or inspecting parts, but we usually do bring some information. For example, if we are instructing a new buyer how to enter a purchase order, we would need to know the product, applicable specifications, quantity, date required, supplier, quotation number, and so forth. This means we should include a field for "Information."

No.

JOB BREAKDOWN SHEET

Operation: Gathering parts
System: Oracle
Information: Item number; desired location
Technical terms:
Common key points:

ADVANCING STEPS	KEY POINTS	REASONS
A logical segment of the operation when something happens to advance the work	Anything in a step that might— 1. Make or break the job 2. Injure the worker 3. Make the work easier to do, i.e. "knack," "trick," special timing, bit of special information	Reasons for the key points
1. Find availability	1. Expand inv. Icon 2. Open S&D 3. Item # 4. Copy & paste item #	1. To access inventory data 2. To access shipping locations 3. To select item # 4. To find quantity available
2. Find location	1. Open quantity on hand 2. Open detail 3. Enter item #	1. To see how many locations 2. To see locations 3. To find locations for item #
3. Move item in system	1. Select inventory transfer 2. Item # 3. From and to locations	1. To engage the move 2. To move the specific item 3. To finalize the move
4. Save	1. Verify screen change	1. To confirm parts moved in system
WHAT	HOW	WHY

Figure 3.5 JBS for gathering parts.

The Advancing Steps and Key Points are best explained with an example (see Figure 3.5). The definition of an Advancing Step stated so far is an action that advances the work. The check question is, "Would the job stop if this action were not done?" When we use a computer, every keystroke advances the job because pressing keys is what is required to do the job. If we used that definition for an Advancing Step, we would have a long list of keystrokes, which would have two main problems. First, the list would be long and thus difficult to remember. This would probably evolve into creating lists of steps for each operation, which would, at a minimum, slow down performing the job. Second, the learner would not understand what is being done so when a similar job arose, a new list must be created and there would be no gains made from accumulated knowledge. In software jobs, we often go from screen to screen and each screen offers us options. When we learn a given job, we usually choose only one option in a screen, although we view the entire screen. When we are being trained for a new job, we may be brought back to screens that we used for another job. Because we have used some functions on this screen in the previous job, we have some familiarity with the screen. We will be using the same screen for the new job, but we will be taught different functions on that screen. As a result, the instruction and retention time may be shortened. The point is that when we are brought to a screen, we will be instructed only on those options or functions that are necessary to do the job in question.

An Advancing Step in a software job therefore mainly describes WHAT we want to do while advancing the work is a secondary criterion. The scenario in the sample JBS—Gathering Parts—is that of a stockroom worker who is moving parts in the warehouse inventory system. WHAT he does first is to find out how many parts are to be moved or the availability. HOW he does this is to click on the inventory icon to expand it so he can see locations. Then he enters the item number about which he is concerned so he can find out how many are available. WHAT he does second is to find the locations where the parts are and locations that are open for new stock. HOW this is done is again described with the Key Points.

Using a JBS for computer applications really simplifies them. In this case, the stock handler is told he must

1. Find available stock
2. Find locations where the stock is and where it can go
3. Move the stock from where it is to where he wants it
4. Save the operation

This is much clearer than having to remember a series of keystrokes. Note that the last Advancing Step is to "Save." If these steps are not saved and the learner goes to another screen, all input would be lost and the job would stop or be incomplete. As previously stated, there are many actions that fall into two or three of the categories (manual, visual, software), and so we must consider all three views when determining Advancing Steps.

Training Aids

There are some jobs that include so many Key Points that it is unwise to attempt to commit them to memory. A classic example is pre-operation checking of a forklift. Most people in industry are aware of these machines since they are often used where materials are being moved. They are a fairly involved piece of equipment, including mechanical, electrical, electronic, hydraulic, and pneumatic systems. Furthermore, most are used on a continuous basis and system failures can cause real harm. As a result, many forklift operators are required to inspect their vehicle before they use it and, as a minimum, at the beginning of each shift. Do an Internet search for "forklift inspection checklist" and you will receive thousands of hits. A common factor among all of them is that there are always 20–30 items to check with each inspection. These are all Advancing Steps,[23] but one cannot be expected to remember all of them on a continuous basis. The result is that we create a Training Aid, which is a checklist of all the items. There are two main ideas to remember about Training Aids. The first is that they should be used only if the most experienced person doing the job uses one. The objective of JIT is to get a person up to production speed as fast as that person is capable. Thus, if an experienced person refers to a schematic, a set of specifications, a checklist, or similar notes or documents in order to do the job, then they should be included in the JBS and the learner should use them also. However, if the experienced person does not use any such aid, then it should not be included in the JBS. If the learner is given an aid that the expert does not need, it will take the learner longer to reach proficiency (if at all) and the aid will become a "crutch."' This, of course, also includes the JBS itself. The second idea about Training Aids is that if it is a checklist, the learner should be taught how to use it. A checklist exists because there are too many items to reliably remember. Some people believe they can save time by checking some or all of the items and then filling out the checklist all at once. Because checklists contain Key Points, omitting one or more at any one time may not cause a problem. However, sooner or later, this expediency will result in a failure.[24] The correct way to use a checklist is to review one item at a time and check it off before proceeding to the next item. Although this may take longer, it insures the quality and safety intended.

Critiquing a Job Breakdown Sheet

It may be easier to critique a JBS than it is to write one from the beginning. You must keep in mind that your objective is to write pertinent notes that the instructor will need to transfer all the information to the learner. You are not writing a standard operating procedure (SOP) or a Work Instruction (WI), so you do not need complete sentences or even articles such as "a" or "the." We want to make the notes as easy as possible for the instructor to grasp and pass on to the learner. If we give the instructor too many words, he will have to paraphrase them for himself and that can degrade the instruction and lead to non-standardization.

Refer to the JBS in Figure 3.6 titled "Placing two factory seconds stickers on a sleeve." Compare this JBS to a revised version in Figure 3.7 titled "Packing factory seconds." The situation is that a product (cookies, batteries, etc.) is coming off an assembly line packed into tubes and two tubes will be packed into a sleeve for sale. A discrepancy has been found in some before they have been placed into the sleeves. One of the duties of the person who has found the discrepancy is to pack them into sleeves with two tubes to a sleeve. They will then be sent to the factory store where they will be sold as "factory seconds." The job entails packing two tubes into a sleeve and sealing each end of the sleeve with a "Factory Seconds" label. The labels are printed as needed with the date, and the person must add the SKU number, the defect, and his or her employee number.

The first item to observe is the name of the Operation. In this case, it's "Placing two factory seconds stickers on a sleeve." Although that is what the person is doing, a better description might be "Packing factory seconds." Applying stickers is just part of what the person does. We want the learner to better understand the scope of the job and not just some of the main points.

The "Parts" will be the product we are making, while the tools are what we use to help us with the job, and the materials might or might not get shipped with the product but definitely are not part of the product. It is not catastrophic if packaging materials are included in parts, but the customer is not really buying packaging.

The next item to review will be the Advancing Steps. In scanning each one, we look for the amount of words, articles ("the," "a," etc.), clauses, which approach a complete sentence, if the action really is WHAT is being done and if it advances the work. A final check would be if we have to tell the learner that action; that is, if the action is simple enough to be obvious. The Advancing Step "open sleeve" is concise and it does explain WHAT is being done. We must open the sleeve in order to pack the product, so it does advance the work. The question to ask is, "Do we have to tell the learner to open the sleeve?" It should be obvious and since we will be demonstrating the job, the answer probably is no.

The second Advancing Step is "Fold Sleeve flaps on one end of sleeve." The number of words should bring added scrutiny. When the sleeve is open,

No.

JOB BREAKDOWN SHEET

Operation: Placing two factory seconds stickers on a sleeve
Parts: Sleeve, factory seconds label
Tools & materials: M/A

ADVANCING STEPS	KEY POINTS	REASONS
A logical segment of the operation when something happens to advance the work	Anything in a step that might— 1. Make or break the job 2. Injure the worker 3. Make the work easier to do, i.e. "knack," "trick," special timing, bit of special information	Reasons for the key points
1. Open sleeve	None	None
2. Fold sleeve flaps on one end of sleeve	1. Fold small side flaps inward on both sides 2. Fold large flap over small side flaps 3. Fold narrow edge of large flap into open edge of sleeve, ensure large flap locks in	Small flaps must be folded first or large flaps will not lock once closed and product could fall out once handled
3. Record data on factory seconds labels	1. Care must be taken to record the proper: • SKU • Defect • And your employee number on each Factory Seconds label	1. All factory seconds are removed from inventory to maintain accurate inventory data 2. Employees buying product want to know the reason it's a factory second 3. Your employee number allows management to provide feedback if you are packing a product that is acceptable for factory firsts
4. Apply "first" of two factory seconds labels	1. Apply first FS label as to position the words "Factory Seconds" (on the label) to the small end of the sleeve 2. Take care not to cover the small SKU printed on the small end of the sleeve 3. Neatly apply the remainder of the sticker over the edge and onto the body of the sleeve	
5. Pack two tubes of product into sleeve		
6. Fold sleeve flaps on open end of sleeve	1. Fold small side flaps inward on both sides 2. Fold large flap over small side flaps 3. Fold narrow edge of large flap into open edge of sleeve, ensure large flap locks in	Small flaps must be folded first or large flaps will not lock once closed and product could fall out once handled
7. Apply "second" of two factory seconds labels	1. Apply second FS Label as to position the words "Factory Seconds" (on the label) to the small end of the sleeve 2. Take care not to cover the small SKU printed on the small end of the sleeve 3. Neatly apply the remainder of the sticker over the edge and onto the body of the sleeve	
WHAT	HOW	WHY

Figure 3.6   JBS for factory seconds (first).

No.

JOB BREAKDOWN SHEET

Operation: Packing Factory Seconds
Parts: Product
Tools & Materials: (2) Factory seconds labels (doc. 4627); sleeve (# 960-9); pen

ADVANCING STEPS	KEY POINTS	REASONS
A logical segment of the operation when something happens to advance the work	Anything in a step that might— 1. Make or break the job 2. Injure the worker 3. Make the work easier to do, i.e. "knack," "trick," special timing, bit of special information	Reasons for the key points
1. Close end	1. Side flaps first 2. Lock large flap 3. SKU 4. Defect 5. Employee # 6. Legible 7. Small SKU visible 8. Smooth	1. Enable large flap to lock 2. Retains product 3. Traceability 4. Customer knows defect 5. Traceability-decision for seconds 6. Transfer information 7. Verify product 8. Presentation/readability
2. Load product		
3. Close end	1. Side flaps first 2. Lock large flap 3. SKU 4. Defect 5. Employee # 6. Legible 7. Small SKU visible 8. Smooth	1. Enable large flap to lock 2. Retains product 3. Traceability 4. Customer knows defect 5. Traceability-decision for seconds 6. Transfer information 7. Verify product 8. Presentation/readability
WHAT	HOW	WHY

Figure 3.7 JBS for factory seconds (second).

it is a rectangular cylinder, so it has two ends. We are told to close one end, but we are not told which end to close first. Thus, the step could be reduced to "Fold flaps." We are actually doing more than just folding the flaps of the sleeve. We are closing the end of the sleeve, so perhaps we should call this step "Close end." We must also ask if closing the end advances the work. The answer would be "yes" since if we did not close the end, the product would quickly fall out. In addition, a stack of sleeves full of product with open ends would be obvious and most likely would not leave the station, so the job would stop.

Advancing Step 3 is "Record data on factory seconds label." Again, many words should bring scrutiny. We ask WHAT is being done and the answer is "recording data." HOW is it being done? Answer: on the Factory Seconds labels. So this step would become just "Record data," but we also have to ask if it advances the work. The answer probably is no. When questioned sufficiently, we would probably find that two additional events could happen. First, the product could leave this area packed but without labels. There could be a possibility that they might go to shippable inventory. The second scenario is that they get packed and labeled, but without any data. Since these are two mistakes that could happen, they are two possible Key Points. Therefore, what we have seen as Advancing Step 3 is actually a Key Point.

Advancing Step 4 is "Apply 'first' of two Factory Seconds labels." Note that we do not have to tell the learner that this is the first label to apply since he has not applied any yet, so hopefully he knows that. We also do not have to identify them as "Factory Seconds" labels since there is no other kind in this job. "Apply label" is definitely *What* we are doing, but does it advance the work? We found out previously that the product could leave this workstation without a label and the large flap has a "lock" on it. Thus, if we do not apply a label, the job would not stop. Thus it is not an Advancing Step but a Key Point.

Advancing Step 5 is "Pack Product into sleeve." There does not appear to be any other container in which to pack the product, so we should be able to reduce this to "Pack Product" or even "Pack." To some the word "pack" might imply the complete operation of loading and sealing a container. Some might be happier with the word "load" to refer to only putting the product into the sleeve. This wording must be agreed upon by all involved in this JBS.

Advancing Steps 6 and 7 duplicate Advancing Steps 2, 3, and 4, so the same reasoning goes for those two steps. Reviewing what we have done so far leaves us with three steps: close end, load product, close end.

We use a similar technique for the Key Points. One difference between Advancing Steps and Key Points is that there is a possibility that there may be no Key Points. If that is the case, the space should be left blank. As the instructor scans the JBS during the instruction, it takes less time to register a blank space meaning "no Key Points" than it is to read the word "none" and translate that to mean "no Key Points."

The three Key Points in Advancing Step 2 do tell us HOW to close the flaps, but we should ask if the instruction can be simplified. The only real information we are gaining from Key Points 1 and 2 is that the small flaps should be folded first. We do not have to tell the learner to fold the flaps or to fold then inward. Key Point 3 tells us to lock the large flap, so we should leave it at that.

Recording the data and applying the label were determined to be Key Points, so we must decide what information we need to tell the learner. The original JBS listed only one Key Point for recording data when actually there are three. These are to include the SKU#, the defect, and the employee number on the label. A Key Point often associated with filling out labels or forms is legibility, so it was included here. The original Advancing Step 4 said to put the label on the "small end of the sleeve." The sleeve has two ends, two sides and a top and bottom, so the labels should be put on the ends. Again, there is no need to tell the learner that since we will be demonstrating the job. Another Key Point is to make sure the small SKU is visible. Since we will be drawing the learner's attention to that SKU, there is no need to tell him to put the label on the small end. In the original JBS, Key Point 3 in Advancing Step 4 is to "neatly apply ..." Thus, we add a Key Point to the revision saying "smooth." This will be sufficient to remind the instructor to tell the learner to avoid bubbles, wrinkles, and smudges so the label can be read easily. In this case, each Key Point has only one Reason, but if a Key Point had multiple Reasons, they should all be listed. If Key Point 1 had three Reasons, we would list them as 1a, 1b, 1c.

Note that each Key Point has a Reason and the number of each Reason corresponds to the number of the Key Point. This makes it easier for the Instructor to reference and for the Learner to remember. A final note is that this particular JBS has two identical Advancing Steps and thus the Key Points for Advancing Step 3 could read, Same as Advancing Step 1. This tells the instructor that s/he can tell the learner a hint. Although there are 16 Key Points for this job, the learner need remember only 8 of them since they are duplicates.

Finalizing a Job Breakdown Sheet

Get Consensus

At this point we have created a JBS draft because only a few people have been involved. Once this has been done, it is time to get a consensus of the JBS. This may be the most difficult part of the process since everyone must agree on the process, wording, and terminology in the JBS. The consensus process is made easier if all personnel in the facility have experienced the 10-hour JIT program. Everyone will not be writing JBSs or using them, but all personnel should know and understand the JIT method, and this is best done by receiving the 10-hour program. Once everyone understands the theory and practice of JIT, determining the best way to perform a given job is much easier. "Everyone" will include any operator who has done the job, group leaders, supervisors, engineering, quality control and inspection, maintenance, human relations and safety, and any others that might be relevant. Each of these groups do not have to be involved in every JBS but only the ones that have meaning to them. It should be obvious why production, engineering, and quality are involved. In that involvement, however, we are not restricting the participation to just one individual since everyone has ideas and should know about various operations. Maintenance would be involved, for example, when equipment is involved. Maintenance would be writing JBSs of their own, especially if they are involved in Total Productive Maintenance. Human Relations and Safety would be involved, especially for ergonomics and hazards. Safety is an integral requirement of a Key Point, so all hazards should be listed. Companies get consensus in a variety of ways and should use whichever method works best for them. It is important to keep in mind the intent to avoid creating a bureaucracy that would result in long delays for consensus. The process should be nimble and quick to react. This is the same system that will be used when someone initiates a new idea for an existing JBS, requiring it to be changed. Getting consensus may require going through some revisions but once that has been done, the new JBS should be pilot tested on a novice. If an existing JBS is revised, it is not necessary to verify with a novice because the JBS has already proven itself. We first check the new JBS with a novice because up to this point only "experts" have seen it. If you can train a novice (who has the required prerequisite knowledge), then it is a workable JBS. Once a novice has been successfully instructed, it should be put into the system. At that time the designated instructors can rehearse their delivery so the resulting instruction is as good as it can be. Refer to Figure 3.8 for a schematic of this process.

Cascading

When creating a Job Breakdown Sheet (JBS), the definition we use to determine what job to consider is: "the activities that a person can understand and remember during one instruction session." *That limits us to only a minute or so of work* and thus most people's work is reflected by many JBSs. These JBSs form a hierarchy with the broadest activity at the top cascading down to single JBSs.

Let's look at the first major job to which JIT was applied—training lens grinders.[25] A competent lens grinder was proficient in 20 skills, which are

74 ◾ *The TWI Facilitator's Guide: How to Use the TWI Programs Successfully*

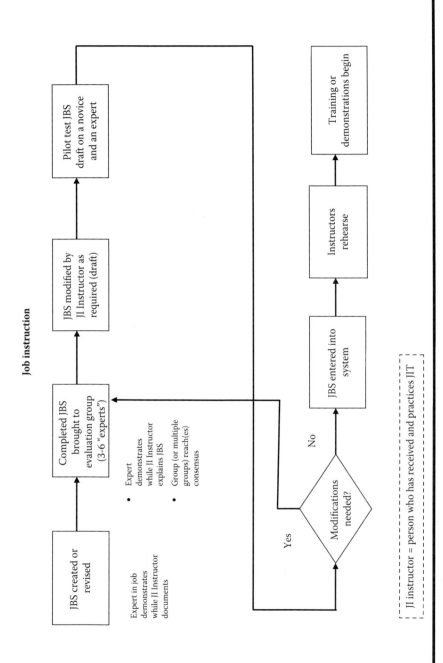

Figure 3.8 JBS creation process.

BOX 3.1 TWENTY SKILLS REQUIRED OF A FULLY COMPETENT LENS GRINDER

1. Cut optical glass
2. Grind lenses
3. Grind prisms
4. Grind reticles
5. Grind windows or covers
6. Correct prisms for polish
7. Blocking prisms
8. Blocking reticles
9. Silvering (ordinary)
10. Silvering (oculars and cutting)
11. Etching
12. Etching (general)
13. Polishing (prism blocks)
14. Polishing small lenses
15. Polishing large lenses
16. Polishing repairs
17. Centering
18. Cementing (lens)
19. Cementing (ocular prisms)
20. Roof prisms (correction)

listed in Box 3.1. Although this is good information to know, it is not useful in training someone to be a lens grinder because merely cutting glass or etching glass doesn't make a product for us. The list of 20 skills can be considered "knowledge," that is, it is something we should know, but is not directly useful to getting the product out the door. In order to be productive, we have to determine what product we want the lens grinder to make.[26] Let's say it's the lenses for an M-1 Circle Aiming Instrument. Refer to Figures 3.9a and 3.9b.

In the context of creating JBSs, you would approach a lens grinder and say that you want to create a JBS for making the lenses in an M-1 Circle Aiming Instrument (M-1 CAI). She would tell you what she has to do to make an M-1 CAI and you would have determined it involves fourteen processes (see Figure 3.10).. When you look at each process, however, you discover that each contains more than just one activity. For example, the fifth

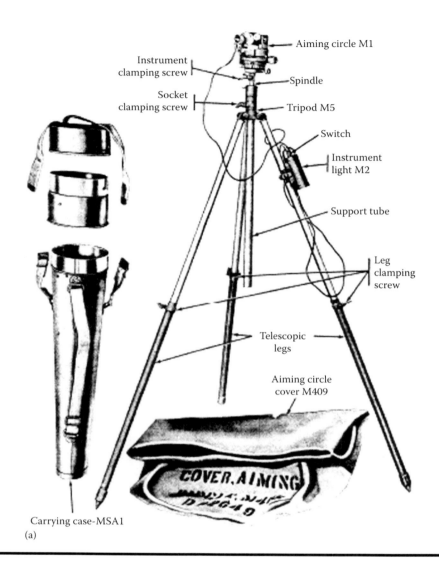

Figure 3.9 **(a) Circle aiming instrument: front view.**

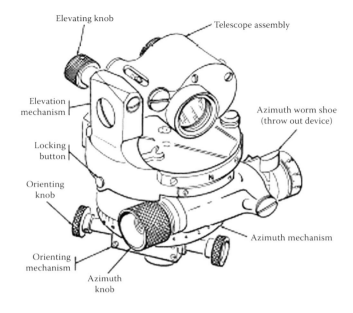

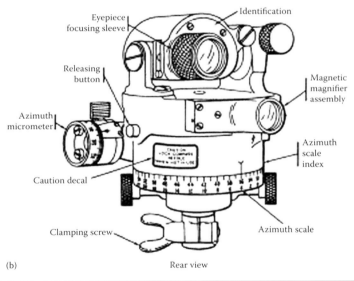

Figure 3.9 (Continued) **(b) Aiming circle M1: rear view.**

process in making an M-1 CAI is to grind a poro prism and that in itself has 14 operations. Furthermore, Operation 6 of Grinding a poro prism (grind two 90° angles) also has six steps. When you look at Step 6, you see that its steps actually are individual activities and thus you can write a conventional JBS for it. Thus, the "job" of making an M-1 CAI cascaded down to a basic JBS of grinding two 90° angles. A JBS therefore must be created for each

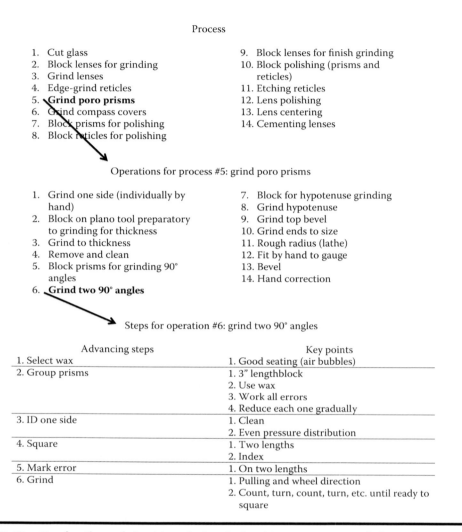

Figure 3.10 Production of the M-1 circle aiming instrument (consisting of poro lenses, lenses and reticles).

base activity. The training then would be conducted by delivering the easiest JBS first and then following with the next most difficult, and continuing until all are covered. Do not follow the JBSs in the order of production; *deliver the JBSs in the order of easiest to hardest.* This method gives the learner confidence and makes her familiar with the job environment. The instruction would start with training someone to grind two 90° angles and continue until all the operations for Grinding a Poro Prism were mastered. Then another of the 14 processes would be chosen and the training pattern would repeat. This would continue until each process of making lenses for the M-1 CAI was mastered. This pattern of instruction reduced the training time for lens grinders from 5 years to 4–6 months, which was ultimately reduced to 6 weeks in the summer of 1945.[27]

Knowing when to write a single JBS and when to cascade actions into multiple JBSs requires some experience. The main concept to be used is to limit the amount of information in a JBS to what a person can absorb (understand and remember) during one training session. If a learner has difficulty because there is too much information, that is a signal that the job should be broken down into more detail. A rule of thumb is to restrict an Advancing Step to one action. Referring to the JBS of changing a car tire, discussed previously, the first Advancing Step is "Raise car." That step includes opening the trunk, removing the jack, positioning the jack under the car, and operating the jack. In some cases, positioning the jack may be involved to the extent that it requires its own JBS. If that were the case, this would be an example of a cascading job. The instruction would begin with the JBS of using the jack, which would consist of obtaining it, positioning it, using it, removing it, and replacing it. Once the learner knew how to use the jack, we could then instruct him how to change a tire. If our method included the use of a torque wrench, there might be another JBS for that. Refer to Appendix II for a cascading example where a series of JBSs is developed for receiving oil from a tanker truck.

Delivering Instruction Using the JIT Method

People absorb information by seeing and hearing. They verify and remember that information by performing a task: that is by doing. The JIT method is different from much other training because it uses a structured approach. The auditory structure is created by the JBS and used in Step 2—Present the Operation. Although the JBS is not a script to be read, it does provide consistent notes for the instructor to use so that any instructor will deliver the same basic information. The paradox is that, when done correctly, it does not sound structured. The only way someone can tell that it is structured is if they hear the same job taught to another person. One can immediately recognize that the same information has been delivered, even though all the words are not quite identical.[28] Kinesthetic learning (doing) is addressed in Step 3 of the Four-Step Process—Try Out Performance. Visual learning is addressed in Step 2 when we present the job to be learned, but it should not be restricted to just that part of the training. Visual cues are important and the concept has been expanded to a body of knowledge referred to as visual workplace management. The underlying concept is that it is more efficient and effective for someone to understand something by seeing it than by asking someone about it.[29]

As mentioned previously, preparation is key, which means that the workplace must be set exactly as it should be. The correct materials and tools must be present. Depending on the job, it is often helpful to have samples available to show the learner exactly what is meant in certain cases. For example, it may be helpful to show the next stage of the product or the final product so the learner can see where his operation fits into the overall scheme. If there are any inspection Key Points, it is useful to show what is expected and what should be avoided. It is not enough to say, "the edge must be free of burrs." Showing both an acceptable edge and a discrepant edge will leave an image in the learner's mind that will facilitate getting the required quality. It is also very beneficial if the instructor has a deep knowledge of the job that goes beyond the JBS. A question may be asked that, although it is relevant to the job, does not directly affect being able to do the job well. If the instructor does not know the answer, s/he should record the question, tell the learner that s/he will get back to him, and then do just that. Asking questions should be encouraged and if people's questions are not answered, they will soon stop asking them. I will briefly describe the four steps of the JIT method.[30] Refer to Box 3.2.

JIT Four-Step Method

Step 1: Prepare the Worker

There are four main points to address in Step 1. First we must put the person at ease because every learner may be nervous to some extent and anxiety will interfere with learning. During all four steps, the instructor should be very aware of the learner's body language and much can be found out during this first contact. The instructor can find out if the learner is outgoing or quiet, knowledgeable or a novice. Talking about something unrelated to work is often done to put the person at ease. People like to talk about what they know, so the instructor can ask a few questions about hobbies, weekend activities and so forth. This will depend greatly on how well the instructor knows the learner and should not take longer than a minute or two. Depending on who the learner is, the instructor might tell him about the department, about the training procedure, or anything else that might be helpful. No matter who the learner is, the instructor should

BOX 3.2 HOW TO GET READY TO INSTRUCT

Create a Timetable

How many skills you expect specific employees to have by what date

Break Down the Job

List important steps

Pick out key points

(Safety is always a key point)

Prepare Everything

The right equipment, material and supplies

Properly Arrange the Workplace

Just as the worker will be expected to keep it

KEEP THIS CARD HANDY

JOB INSTRUCTION

TRAINING

www.TWIPartners.com

How to Instruct

Step 1: PREPARE THE WORKER

- Put the person at ease.
- State the job and find out what the person already knows about it.
- Get the person interested in learning the job.
- Place the person in a correct position.

Step 2: PRESENT THE OPERATION

- Tell, show, and illustrate one *IMPORTANT STEP* at a time. Present in Chunks.
- Stress each *KEY POINT and REASON*.
- Instruct clearly, completely, and patiently, but do not give more information than the person can master.

Step 3: TRY OUT PERFORMANCE

- Have the person do the job: Correct errors.
- Have the person repeat the job and explain each *IMPORTANT STEP, KEY POINT, AND REASON*.
- Make sure the person understands.
- Continue until YOU know THE PERSON knows.

Step 4: FOLLOW UP

- Put the person on their own. Designate to whom they go for help.
- Check frequently. Encourage questions.
- Taper off extra coaching and close follow-up.

If the worker hasn't learned, the instructor hasn't taught.

make sure that s/he knows where s/he and the job s/he is about to learn fit into the overall company strategy. This is a good opportunity to make sure the learner knows the training procedure. Many people are nervous when being instructed because much of their previous experience with instruction has been stressful. Tell the learner that you will perform the job several times and then you will watch him or her perform it several times. Avoid saying how many repetitions will be done because the learner might think that s/he has to master it in that number of times. The learner should understand that you will work with him or her until they can do the job. Of course, once JIT is a well-known method in your organization, this will be understood. In fact, once JIT is part of the culture, employees will not accept any lesser quality of instruction. Sometimes the person will want to perform the job as you are demonstrating it or take notes during the instruction. These are methods learners use to help compensate for poor instruction. Explain to them that the JIT method is such that those actions are not necessary and actually counterproductive. If the learner does the job while the instructor does it, the learner cannot watch the instructor. The learner will be given an opportunity to do the job, but first we want to make sure s/he knows what to do. Similarly, taking notes during instruction interrupts and therefore decreases the flow of knowledge. If the learner cannot be dissuaded from taking notes, there will be two requirements. First, when the learner wants to take notes, s/he must tell the instructor to stop because s/he cannot take notes and observe at the same time. Second, when the learner demonstrates the job for the instructor, s/he will not be allowed to use the notes.[31]

The second main point in Step 1 is to find out the person's knowledge that is relevant to the job. Most people do not make industrial products at home, so the transfer of knowledge will not be exact. However, if the job requires handling small parts and the learner builds scale models, s/he would be used to handling small parts. If the job entails assembly or disassembly and a person uses a sewing machine at home, they may be used to taking equipment apart since they must change the bobbin and do first-level maintenance and/or adjustments on the sewing machine. Obtaining this information is important because we do not want to talk down to the learner just as we do not want to "talk over his head." In either case, we'll "lose" him and the instruction will not go well. This also gives the instructor an opportunity to find out if the learner has the prerequisite knowledge for the job. For example, if the job requires the use of an instrument or tool, the instructor should inquire whether the learner knows how to use the instrument or tool. If the answer is "no," the training should stop and the learner should be trained on using the tool in question.

The third point is to tell the learner about the job so that the learner knows where it fits into the overall product. The real key here is getting the learner to understand why this particular job is important to the total production. Every job is important or we would not pay people to do it, but sometimes the importance of a job is subtle. People are paid to clean offices and shops. They are either the workers themselves or people whose only job it is to clean. In either event, they would not be paid to clean unless it were important. When we are instructing someone on cleaning, we often do not tell him or her why it is important, leading them to find pride in the job themselves. Some people will do this but most will not. *If we want people to care about what they do, they must know why the job is important and we must tell them.*

The fourth point is to make sure the person can see whatever it is you are doing. That may sound obvious, but many people demonstrate and assume the learner is seeing what they are supposed to see. Make sure the actions are visible to the learner whether it is happening on a table, bench, or computer screen.

The above points do not take a long time, but they will result in improved instruction. Once we are satisfied with Step 1, we go to Step 2.

Step 2: Present the Operation

The main Key Point in presenting the operation is *to give the learner a little information at a time.* The pace will depend on the learner's existing knowledge, which you have already discovered to some extent. You must have the JBS available so that you can glance at it occasionally to make sure you have not left out anything or put anything out of order. The JBS will assist you in what words to use so that we can get standard work. For example, in the Fire Underwriter's Knot, the first Advancing Step is "untwist" and the instructor should say "untwist" and not "unwind," "unravel," "uncoil," "undo," or any other synonym. Using different words for the same object or action can cause confusion in an organization, so the main words for Advancing Steps and Key Points should be standardized. The JBS helps us do this. The job should be done several times, which means at least two,

84 ◾ *The TWI Facilitator's Guide: How to Use the TWI Programs Successfully*

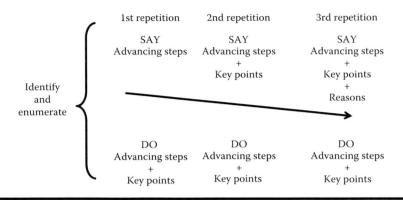

Figure 3.11 Visual aid for delivering instruction (Step 2).

usually three, and as many times as the learner requires before he tries it himself. The first iteration consists of saying just the Advancing Steps. The second iteration consists of saying the Advancing Steps and the Key Points. The third iteration consists of saying the Advancing Steps, the Key Points, and the Reasons. Refer to Figure 3.11 for a schematic visual aid describing this pattern. In Step 1, Prepare the Learner, there was no limit to what you could say. The learner must only remember concepts and even if they are forgotten, no great damage has been done. Step 2, Present the Operation, however, contains the heart of the job and you want the learner to understand and remember everything you say. To help him out, you must use "JIT Speak."

To help the learner understand the job you must identify whether you are telling him an Advancing Step or a Key Point. Since both are actions, it will be difficult for the learner to quickly tell. In addition, an action can be an Advancing Step in one job and that same action can be a Key Point in another job. An example of this was already seen in the JBS of Load Hardware Bag (Figure 3.3). The action "attach wire tie" is a Key Point when the tote is used, but it is an Advancing Step when the shipping pallet is used. Another example of this is when we check something. Say the action is "Check the bolt length." If that action appears in an assembly job and nothing will be done with the bolt length at this station, it is a Key Point. The reason to check the length might be to make sure it is correct for future operations. If we do not check the length and it is incorrect, our job will not stop. A future job several workstations after us will find the problem, but we will have done our job (although incorrectly). Thus, we will have caused scrap or rework even though we were apparently successful in what we were doing. However, if "Check bolt length" occurs in an inspection job where we have to check several parameters, each one of those parameters would be an Advancing Step. If any of them were not done, the inspection job would stop. This is why separate inspection jobs are inserted into some operations. When inspection is all that you do, it is easier to keep track of each step. This is also why the dichotomy of Advancing Steps and Key Points helps us reduce the amount of errors.

To help the learner remember the Advancing Steps and Key Points, we should number them. Whenever we want to remember a list of items, it is always easier if we number them. If you go to the grocery store to buy five items, you may think you do not need to write a list. When you are at the store and have bought three or four of these items, you may return home if you have purchased those you remember best. If you know you were to purchase five items and you have picked up only four so far, you will stop and try to remember what the last item was. Numbering items is very important when there are several items we must remember, but it is useful to number all Advancing Steps and Key Points so that we are in the habit of always stating the number.

Identifying and enumerating Advancing Steps and Key Points results in a script that will be very close to the following. (Refer to Figure 3.2.)

> All dialogue by the Instructor:
> (First iteration)
> "Now I am going to show you how to change the tire on a car.
> There are five Advancing Steps. The first Advancing Step is to raise the car."
> The instructor silently steps away from the car so that he can glance at it from a distance. He then gets into the car and checks that the parking brake is engaged and that the car is either in "park" or in gear. He then goes to the trunk, gets a wheel chock[32] from the trunk and places it behind a wheel that will not be changed. He takes the tire pressure gauge from its storage place, opens the trunk, and checks the air pressure in the spare tire. He then takes the tire iron and the jack, walks to the tire to be replaced, and removes the hubcap. He loosens the lug nuts and positions the jack. He proceeds to raise the car.
> Once the car has been raised, he says,
> "The second Advancing Step is to remove the lug nuts."
> He then silently removes all the lug nuts and places them into the upturned hubcap.

He then says, "The third Advancing Step is to remove the tire." He then silently removes the tire and takes it to the trunk of the car.

There are three main concepts[33] in what the Instructor did. He identified and numbered the Advancing Steps. Anyone listening would clearly know what the first three Advancing Steps are. In addition, he limited his words to just saying what they were. He did not talk about Key Points or any other facets of the Advancing Steps; he remained silent. This is very difficult for many people to do. We dislike silence and feel more comfortable while we are talking. We know a lot about this job and we want to tell the learner everything we know. The fact is that we have spent some time analyzing the job and have decided exactly what is important to tell the learner. Thus, nothing else is necessary. The more we say that is not necessary, the more we dilute the ability to learn. The first Advancing Step has two words and we used nine words to describe it, but the learner should be able to remember those nine words.

The very difficult part of this first iteration is that the first Advancing Step contains nine Key Points and they all must be done without talking about them. The learner has enough to absorb just by watching all the actions. He knows the "title" of these actions is to raise the car, but he is probably creating questions as he watches everything the instructor is doing. Every time the instructor does the job, he should do it correctly, which means he must perform all the Key Points. However, in the first iteration, he says <u>only</u> the Advancing Steps. Saying more would overload the learner. Since this is a fairly long job, when the instructor completes the first iteration, he may summarize it by repeating the Advancing Steps. This gives the learner a concise list of steps that are easy to remember.

1. Raise car
2. Remove nuts
3. Remove tire
4. Replace tire
5. Lower car

The instructor then starts the second iteration and says,
"Now I am going to show you some Key Points for those steps. There are four Common Key Points that pertain to all the steps and they are the first actions you take. The first Common Key Point is to check that the car is on a level surface."

The instructor backs away from the car, glances at it, and nods his head.

"The second Common Key Point is to put the car in 'park.'"

The instructor gets into the car and checks that the car is in "park" or in gear. Note that he would direct the learner to follow him and sit in the passenger's seat so he can see what is being done.

"The fourth Common Key Point is to chock at least one wheel."

The instructor goes to the trunk, gets the wheel chock and places it behind a tire. Again, he makes sure the learner can see what is being done. He could also make a point of showing the chock to the learner before he puts it behind the tire. Note that the instructor will tell the learner the purpose of the wheel chock in the next iteration and not this iteration. The instructor continues:

"There are five Advancing Steps. The first Advancing Step is to raise the car. That Advancing Step has five Key Points. The first Key Point is to check the spare tire's air pressure. You mentioned that you know how to use a tire pressure gauge.[34] Is that correct? The rated pressure is noted on the driver's door jam, in the manual, and sometimes it's on the tire."[35]

The instructor gets the tire pressure gauge and checks the air pressure.

"The second Key Point is to use the blade of the tire iron to remove the hubcap."

The second iteration continues through all the Advancing Steps and all the Key Points. It is important to say, "The first Key Point is …. The second Key Point is … The third Key Point is …" as they are being done. Avoid saying, "The next Key point …" or "The last Key Point …" because this does not tell the learner where you are in the list of Key Points. This is the part of the instruction that takes practice or rehearsal. A check to determine whether "JIT Speak" is being used is if an observer can quickly and clearly write a JBS that matches identically with the one the instructor is using. When the instructor says, "The fourth Advancing Step is to replace the tire. That step has five Key Points and the first Key Point is to clear away any debris," it is very clear what is written on the JBS.

Notice that the instructor used more words during the second iteration to describe and explain the Key Points. During the third iteration, he may use even more words so that he is sure the learner understands the reason for each Key Point. The script would read like this.

"Now I am going to do this job again and tell you the reasons for all the Key Points. There were four Common Key Points. The first Common Key Point was to check to see that the car is on a level surface. The reason we want to do that is to reduce the likelihood of the car falling off the jack. Raising the car with the jack is safe when done on a firm, level surface, so we want to make sure the surface is firm and that the car is level. Be very aware of the stability of the car when using the jack. It should not move or 'wobble' when raised. Never reach under the car when it is raised and stay as far away from the car as possible. If the car does fall off the jack, aside from getting hurt, the car may be damaged and you may need a tow truck to lift the car off the fallen jack."

The point here is that when explaining reasons, you have a platform to explain all the consequences of not doing a Key Point, that is, of making a mistake.[36] The four Common Key Points in this example are safety Key Points, so you want to make an impression on the learner to make sure he follows them all the time. The words listed in the "Reasons" column are "trigger" words to remind the instructor what to say. For example, the reason for checking the air in the spare tire is if it is deflated, the person should not go any further. In this instance, this Key Point has actually stopped (prolonged) the job. The tire pressure, however, is not an absolute, so if it has 27 pounds of pressure when it should have 30 pounds, the learner should know that he could change the tire but he should get it inflated as soon as possible.

Three iterations are usual, but the instructor should never assume that is enough. S/he should ask

> Would you like to try the job now or would you like me to do it again?

If the learner would like the job repeated, the instructor should repeat the job while saying all Advancing Steps, Key Points and Reasons. At the end of that, the above question should be repeated, followed by the same pattern. Once the learner has enough confidence to try the job, proceed to Step 3—Try Out Performance.

Step 3: Try Out Performance

There are two main ideas to remember in Step 3. The first is to make sure that the learner can do the job. We want to see the learner perform the

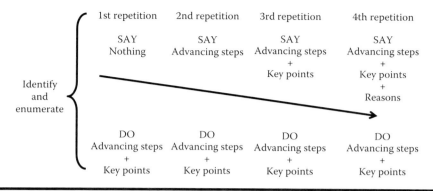

Figure 3.12 Visual aid for performing job (Step 3).

job correctly and without assistance at least one time. Once that has been accomplished, we want to make sure the learner understands what he is doing and that he can explain it to us. The schematic showing this is in Figure 3.12 JBS visual aid 2. When the person is performing the job, watch carefully to be certain that all Key Points are being done. Some Key Points may not be visible, but it is still incumbent on the instructor to make sure the job is done correctly every time it is done. When the person has performed the job once correctly without assistance, we can be fairly certain that he knows how to do the job. We must now determine if he really understands everything about the job.

The instructor should praise the learner for performing the job correctly and then ask him to repeat the operation and identify the Advancing Steps as he does them. Sometimes a learner will say that he does not believe he can name all the steps. The response is that since he did the job once, he should be able to do it again. This time, however, we are merely going to put labels on what he has done. People who are not familiar with the concepts of Advancing Steps and Key Points often mix them up when asked to name them, but the instructor can correct the learner as needed. Once the job has been done this second time, the instructor should praise the learner and ask him to repeat the job and identify the Key Points for each of the Advancing Steps. After this iteration, the job is repeated with the learner telling the Advancing Steps, the Key Points, and the Reasons. At this point, the instructor has seen the learner perform the job correctly four times and it was explained fully. In most cases, the instructor is fairly sure that the learner knows the job well enough to be on his own, but he should always ask the learner. Some learners like to perform the job a few more times just to gain confidence, while others say they are ready to go on their own. In any case, the instructor should follow what the learner says and also make sure that the learner has no further questions. At this point, we enter Step 4—Follow-Up.

Step 4: Follow-Up

Follow-up is always required because, as confident as the learner is, questions have a tendency to arise once the instructor leaves. If the job is highly repetitive, the instructor should return in a few minutes so that the learner has time to do 10–20 repetitions of the job by himself. If the job is longer, such as changing a car's tire, the follow-up would consist of watching the learner the next time the job arises. Alternately, a follow-up session could be scheduled within the next day or two so the learner can demonstrate his knowledge of the job. The follow-up should be gradually reduced so that the learner gains confidence and recognizes that the instructor is actually handing the job to him.

Job Safety Training

The TWI Service was repeatedly asked for special programs in addition to the three main J programs.[37] Its answer was always the same. "You have the tools, go ahead and use them."[38]

Technically, there is no such thing as "safety training." There is only training to do a job safely or unsafely. The TWI Service recognized the dichotomy of knowledge and skills. Knowledge is something we learn by seeing, hearing, or reading; it is something we know. A skill, on the other hand, is something we do and improve with practice. Safety falls under the heading of knowledge. Companies require "safety training" and comply with regulations by gathering employees into a room, lecturing them for an hour or more, and recording attendance. This more accurately should be referred to as "acquiring safety knowledge." However, since the participants are only seeing and hearing the information, but not doing anything, what the participants actually absorb is questionable. The best result is that the employees now know actions that are safe and unsafe. Even if they pass a written test, there is no certainty that they will employ those actions on the job because they have not actually used them. In order to be safe or unsafe, a person has to do something. Thus, Job Safety Training is actually a subset of Job Instruction Training. This means that when JIT is correctly implemented, the use of "safety training" as is commonly done should be discontinued. This does not mean we are discounting the importance of safety. To the contrary, because safety is so important we want to address it in the most effective way possible and do not want to diminish its importance by conducting sessions that are not productive and in which people do not see the benefit.

When we are instructing someone in how to do a job, we list all the safety Key Points. Safety, by definition, is always a Key Point and can never be an Advancing Step because safe practices, by themselves, do not advance the job. If they did, we would not have to remind people about them. The main intent of Job Safety Training is to write a better and more complete safety Key Point by considering all possible hazards and possibilities. The only way to eliminate a Key Point (including a safety Key Point) is to mistake-proof the job. When robots started being used on the manufacturing floor, cages were built around them so people would not walk through its operating area. Robots would not sense people and someone could easily get hurt if they were not careful. The procedure maintenance technicians were supposed to use when working on robots was to shut off the power to the robot before entering the area. This is a Key Point. Naturally, this did not always happen and after some injuries, "kill switches" were put on the doors to the areas so that the power would automatically be turned off whenever the door was opened. The robot would have to be started manually once the door was shut. Every hazard is not as obvious as that and every solution is not as easy as that. For example:

- Wood skids have splinters.
- A dull knife will cut your finger before it cuts the box.
- That carrier does not belong there and it's so low it's a tripping hazard.
- It will only take me 5 seconds to empty the solvent, so why do I have to put on a face shield and gloves?

When safety training is done separately from production training, people will think about it separately. That is acceptable *if* they think about all aspects of the job. When tired or in a hurry, people have a tendency to remember the production steps and not the safety Key Points. We want people to think that the safe way is the only way to do the job. Safety is extremely important and everyone should be continuously aware of hazards. However, the hazards will vary with the workplace and it is unreasonable to expect everyone to be constantly aware of all hazards. What is more reasonable is for people to be aware of hazards in their own workplace and in their own jobs so they react to them appropriately. The safety committee should be aware of all hazards at a given facility and that is why they should review each JBS before it is used. This would be done in the process of getting consensus. They can then make sure that all appropriate hazards are accounted for.

JIT beyond Manufacturing

The JIT method is often associated with the manufacturing shop floor because that is where many examples originate. In addition, mechanical or hardware jobs are easier to envision. However, it should be remembered that the TWI Service applied JIT to all defense contractors and that included the entire supply chains for production of equipment, clothing, and food. That would include the purchasing and accounting departments, and also hospitals, where it is being used today. Whenever you do something, you are using a process, even if you haven't written down the process. W. Edwards Deming said, "If you can't describe what you are doing in a process, you don't know what you're doing." When you describe your job as a process, you can create a JBS for it. JIT can also be used for instructing people how to troubleshoot equipment because that is also a process. There is a certain pattern troubleshooters use when they are analyzing a problem and that can be captured in a JBS.

JIT Creation Process

At this point we have written and gotten a consensus of a JBS. This is the best method for doing the job that we collectively know at this time. The next step is to verify the instruction with a novice. (Refer to Figure 3.8) So far, the only people who have any input are the people who know the job well. When we instruct someone who knows nothing about the job (and has the correct attitude, aptitude, and prerequisite knowledge) we may gain some insights with respect to how most people will receive the instruction. The biggest hurdle in getting consensus is agreeing on what words to use.

Novices can often help us with this because they have no biases about the job in question. If suggestions are made or questions that we cannot answer are asked, we return to the group who gave the final approval for more information. This may not happen frequently, but it is a good check. Once that has been done, we are now ready to enter it into the system and select instructors. The instructors should rehearse the delivery so that it sounds smooth. Less time will be spent rehearsing the delivery once instructors gain experience, but initially the requirement of saying "There are five Advancing Steps. The first Advancing Step is …" will not come naturally. The trainer and other instructors should watch the rehearsal so they can give feedback and let the instructor know when s/he is ready. Once it is deemed that the instructor is ready, the operators in the selected area should all be trained using the JBS. The trainer should watch the first few instructions to verify that the instructor is on track.

JIT Auditing and Updating

If an auditing or assessment system has not been made by this time, one should be made now. Auditing is necessary because people who have done the job before will probably have to change their habits and people new to the job must create new habits. A properly designed and well-executed audit program is vital to maintaining the quality and the results of the JIT program. It is much like weeding a garden. You can prepare the soil, select what plants you want to grow, and even add pre-emergence material and ground covers to prevent undesirable plants from growing. Because the wind carries seeds, some of which are very tenacious, you must periodically check your garden to see that it is really as you designed it. The team driving the TWI Program should decide on how people will be audited. Operators should be audited to make sure they are following the JBS. Instructors should be audited to make sure they are following the JIT method and trainer(s) should be audited to verify that they are following the JIT Trainer's manual. A coach can sit in on a JIT session to confirm that the trainer is following the method and the trainer or a coach can observe instructors periodically to see that they are following the JIT method. Almost anyone who knows how to audit can audit the operators. This would include senior management on an infrequent basis (say once a month), area supervisors more frequently (say once a week), and standard auditors more frequently. The number of times an operator should be audited should be based on how long it is thought that it will take a person to habitually perform the job according to the JBS. Depending on the job, an audit might be conducted three days in a row. If the operator passes all audits, then the frequency could go to once a week for three weeks and then once a month for three months. Perhaps only a yearly audit would be needed after that.

Both the auditor and the person being audited should think of the audit as an educational process and not a punitive one. Some people do not even like the word "audit" because it is thought of as punishment to some. During the instruction, the instructor should have explained to the learner that the JBS reflects the best way to do the job that is known at the present time. The instructor should obtain recognition of that from the learner so that they both know that the learner will do his best to follow the JBS. In addition, the learner should be told during the instruction that he will be audited at various points to help him with that goal. Therefore, the learner will be expecting an audit. If the learner is not following the JBS at some point during the audit, the question to be asked is "Why?" The learner will not be reprimanded but will be corrected, and the reason for the variation from the JBS will be corrected.

There are three reasons someone does not follow a JBS after s/he has been properly instructed with a JBS.

1. The person forgot the contents of the JBS.
2. The person does not agree with the contents of the JBS and
   a. Wants to do the job the "old way."
   b. Wants to do the job a new way.

#1: Depending on how much the person has forgotten, the correction could be just a reminder of a Key Point or a complete re-instruction session. Each situation should be viewed separately, but there are some questions the auditor can consider.

- Is the JBS too large?
- Does the operator have the required aptitude, attitude, and prerequisite knowledge?
- Was this operator given the appropriate JBS?
- Was the instruction done properly?
- Is the interval between instructing and performing appropriate?
- How many times has this operator been audited on this job? If this is not the first audit, what were the results of previous audits?

#2a: If the operator wants to do the job the way s/he used to do it, it must be explained why that method is inferior to the method described on the JBS. Of course, this should have been done during the initial instruction, but there are reasons this discussion may not have made an impression on the operator. Often a Key Point is required to prevent an error that might occur in subsequent operations. If the error is not common and does not occur frequently, and it occurs far away from the operator, s/he may easily dismiss the Key Point as an annoyance. The auditor/instructor must now go into greater detail to explain the repercussions of not performing the Key Point. When using a JBS there is no such thing as a "shortcut" since everything in a JBS is required for a specific reason and nothing is included that does not have a reason.

#2b: If the operator has found a different way to do the job that varies from the JBS, the new method should be reviewed and the operator's action should be encouraged. However, the execution of the action should be corrected. That is, improvements are always welcomed and encouraged, but they must first be vetted to make sure there are no discrepant repercussions. A small change in one operation may have some effects on a subsequent operation of which the initial operator is not aware. It should be emphasized during the initial instruction that the operator is to follow the JBS because the operator knows of no better way to do the job at that time. If s/he thinks of a better method, a standard change procedure should be followed. Advantages of using standard work are that more people will be thinking of ways to improve a specific job (because everyone is doing it the same way) and when a change is adopted, it has a greater effect (because everyone is doing it the same way).

Documentation

A documentation system must also be created since once a JBS has been created, it should be accessible to everyone who is eligible to use it. The uses would be for instruction and auditing. Methods of documentation vary all the way from keeping a hard copy at the workstation to having a digital copy available whenever someone wants to use it. Variations will be based on the size of the organization, the number of separate facilities, the state of the technology and personnel available. Most companies, of course, will have digital copies but access can vary from anyone to only the necessary few. Having a "read-only" access can be helpful since that would put more control over changes while giving everyone access. This implies that a person or group of people would have authority to make changes. A change system must be created and published but it should be kept as simple and informal as possible to encourage as many changes as possible, while at the same time having a robust method of changing processes. The TWI coordinator could be responsible for processing changes by working with a group from engineering (manufacturing and design), quality, maintenance, finance, and human resources. The coordinator could field suggested changes and work with appropriate personnel to have them vetted and accepted. This, of course, would be required even more when JMT is adopted.

Part of the documentation effort would be keeping the training matrix up to date.

Notes

1. Goodreads. http://www.goodreads.com/quotes/119387-if-you-know-the-why-you-can-live-any-how.
2. This is a term used by Frederick Taylor in writing about Scientific Management. This was a person who was perfectly suited to the job at hand and could perform it as well as it had to be done.
3. Takt time is the average time from the start of one unit to the start of the second unit. The idea is to match the production rate with the demand rate so there is no over production and no unfilled demand. It is calculated by dividing the available time for production by demand (units per period). Thus if there are 40 hours available per week and the customer demands 80 units per week, the required Takt time to make the units is 30 minutes. Takt time can be calculated for a final product or any part of a product.
4. YouTube.com. http://www.youtube.com/watch?v=MFzDaBzBlL0.
5. Dietz, J.W. 1970. *Learn By Doing: The Story of Training Within Industry*. Walter Dietz, Summit, NJ, 1970, Foreword.
6. The principle discussed is 1:1 instruction. Often there is a need to instruct several people in the same job, but note that the 1:1 principle still applies. However, it can be acceptable to instruct one person while the others are observing. They will hear some of the instruction and have some of their questions answered. Once the first person has been trained, s/he can practice the job or become an observer while the instructor instructs the second individual. The first person chosen should be the least self-conscious of the group.
7. Dinero, Donald A. 2005. *Training Within Industry: The Foundation of Lean*. Boca Raton, FL: CRC Press, p. 68.

8. http://www.skillstoolbox.com/career-and-education-skills/learning-skills/effective-learning-strategies/chunking/.
9. Yang, G., San Lai, C., Cichon, J., Ma, L., Li, W., Gan, W.-B. Sleep promotes branch-specific foundation of dendritic spines after learning. *Science*, http://www.sciencemag.org, Vol. 344, Issue 6188, June 6 2014, p. 1173.
10. Vince Lombardi; http://www.brainyquote.com.
11. Refer to Figure 3.8 for a schematic of the JBS creation process.
12. The caveat here is that there can be two different JBSs for the same job and both can be correct. The JBSs should be similar, however, and differ in wording or perhaps some Key Points.
13. The worker to whom we refer is the person doing the job. If a person omits a Key Point and someone beyond that operation gets injured because of it, that is a quality problem.
14. A JBS precludes shortcuts because every action stated (Important Steps and Key Points) is required for the job to be done correctly every time it is done. If someone were to create a shortcut in a JBS, s/he would eliminate a Key Point. Depending on what the Key Point is, they may not see the affect of removing it for many iterations of the job. An experienced electrician who doesn't shut off the power may get shocked on only one job in 100.
15. Recording a JBS on a tablet or laptop computer is very effective, but if they are not available, a pencil should be used since many corrections will be made.
16. A manual JBS could be one for an operator and an object, while a relationship JBS would be one for an operator and a person.
17. Note that in this case the product is the information obtained from an interview.
18. Note that "remove lug nuts" is an Advancing Step because the job would stop if it were not done. Attaching lug nuts is a Key Point since the wheel could be mounted without them, albeit poorly.
19. "Special" Key Points (in the column) as opposed to "Common" Key Points (in the header); the reader may believe that Step 3 should have some Key Points such as "proper position" or "lift 25#," but for the purpose of introducing and explaining this concept, the step does not have a Key Point.
20. In Ireland, hotels have signs on internal doors saying, "Fire Door. Keep Closed." This does not stop people from passing through the doorway because they know the meaning is to close the door after they pass through. A sign over a flotation ring next to a swimming pool says, "Safety Device. Do Not Remove." If someone were drowning, I believe no one would hesitate removing the ring and tossing it to the person in the pool.
21. At this point I was still referring to "required actions" as "Important Steps," which is what caused some of the confusion. This experience started me thinking about the words "Important" and "Advancing."
22. "Software" may not be the best word to describe this method. Other suggestions are "computer" and "programming." This application includes jobs such as setting up a DVD player, finding inventory in a stockroom using a computer or creating a spreadsheet on a computer.

23. This is an inspection job, so each major action is an inspection and if any is not done, the inspection cannot be considered complete: the job stops.
24. An example would be checking the oil level in an engine. A gasoline engine can run with little or no oil, but it will heat up and the pistons will freeze when the level is too low. The Key Point is to check the level daily (as recommended by the manufacturer) so oil can be added before the level is too low.
25. Dinero, Donald A. 2005. *Training Within Industry: The Foundation of Lean.* Boca Raton, FL: CRC Press, pp. 92–4.
26. This is an important concept to understand. Instructing a person to perform a task is more relevant to the person than instructing him how to perform a skill. The purpose is more obvious. Since the instruction is more relevant, it will be easier for the learner to understand and remember it.
27. Training Within Industry Service. 1945. "The *Training Within Industry* Report: 1940–1945. Washington, DC: War Manpower Commission Bureau of Training, pp. 18–20.
28. The words used for the Advancing Steps and Key Points will be identical, but the supportive words that help to form the sentences will vary with the instructor.
29. If I'm looking for a file, it is faster for me to scan file cabinets that are clearly marked and organized than it is to find someone, interrupt their work, and ask.
30. These are the steps that each person who has received the 10-hour JI training should have received on a pocket-sized card. These cards were consciously developed during World War II as a method to keep these important concepts in the minds of those using the JIT method.
31. Note that if the organization complies with ISO, any notes must be a controlled document and personal notes are not allowed.
32. A wheel chock is a wedge-shaped block that prevents a wheel from rolling.
33. These concepts are actually Key Points, but I didn't want to use the term here because it might be confused with the Key Points of the tire changing JBS.
34. The same thinking would be done for any of the required tools. Positioning and using the jack may require a JBS of its own and would be an example of cascading.
35. Note that these questions of using a pressure gage and knowing where to find the pressure level could have been asked in Step 1, Prepare the Worker, and/or they could be confirmed here.
36. Remember that a JBS is not a script to be recited as an actor would in a play. The JBS consists of notes to the Instructor to be complete yet succinct.
37. Union Job Relations was a fourth program designed especially for union stewards and it closely matched the JRT program. In addition, of course, there is Program Development, which addressed the needs of knowledge of work and knowledge of responsibility.
38. Training Within Industry Service. 1945. "The *Training Within Industry* Report: 1940–1945." Washington, DC: War Manpower Commission Bureau of Training, p. 264.

Chapter 4

Understanding JMT

> It is not necessary to change. Survival is not mandatory.
>
> **W. Edwards Deming**

Job Methods is a plan to help organizations produce greater quantities of quality products or services in less time by making the best use of personnel, machines, and materials now available.

As with Job Instruction Training (JIT), first we want to understand why we should be using Job Methods Training (JMT). This program teaches people how to vet, sell, and implement their ideas. The intent is to get more ideas (improvements) implemented by involving more employees. This creates an improvement culture, which can be almost self-sustaining. Teaching everyone these skills may very well be the single most important activity of any continual improvement program, and the one that most people do not act on.

Objectives of JMT

If there is a definite, obvious problem with a job or task, we will approach it in a "problem-solving" mode and often come up with a solution. Often, "brainstorming" is part of this method, where we attempt to become creative and solve the problem with new ideas, or at least by looking at the problem in a different way. This method does not always eliminate waste or give us an optimum solution, however. If the problem is not obvious, we may not even recognize it as a problem, and merely "work around" the situation, treating it as a nuisance. These "workarounds" can introduce waste or other problems and complicate what we are doing. Since increasing productivity is such an important factor in running any organization, we should not treat it with random or haphazard methods. There should be a standard method for addressing productivity improvements at all levels of the organization, and by all employees in the organization. The objective of JMT is to implement all applicable and productive ideas of all employees on a continuing basis. This is achieved by using a standard improvement method, which is based on nurturing, in all employees, a questioning attitude. The method must be straightforward enough that all members of the organization can easily understand and use it.

Undoubtedly there will be some employees whose personalities are such that they are very eager to make improvements. Some of these people may even have positions with "improvement" in the title, such as Continual Improvement Manager. Others, who have received training and have some experience in making improvements, may assist this position. These employees may be called Black Belts or Green Belts. This group or department conducts large and small projects and does a credible job of improving many facets of the operation. At the other end of the spectrum are at least as many employees who want no part of improvements or change of any sort. Their belief is that they have been paid to do a job and that is what they intend to do. That, of course, leaves the 80% or so in the middle who are willing to improve their jobs, and have some ideas about how to do it. However, they either do not know how to make the necessary changes, or they are not allowed, or think they are not allowed to do it. The objective of JMT is to reach this population and enable them to implement their ideas. This is done through JMT by instructing everyone, and by using a standard method that everyone can understand and use on a repeatable basis. Since the three groups are not wearing shirts that quickly identify them, we must reach all employees with JMT. The first group should embrace JMT, while we may have to depend on peer pressure to activate the second group, which is at the other end of the spectrum. We do that by using the improvement team to facilitate the implementation of JMT with the 80%.

Benefits of JMT

The direct benefit of using JMT is the gain made by the large number of productivity improvements that will occur throughout the entire organization. JMT is applicable to all facets of an organization, including all office

**BOX 4.1 PROCESS IMPROVEMENT
VERSUS PROBLEM-SOLVING**

JMT is a process improvement program in that it is used to improve a process that is already in place. Someone believes it can be made more effective or efficient, which, of course, is true for any process. What is to be determined is not if the process can be improved, but rather how it could be improved and if the cost in time and/or money is justifiable. A problem-solving program on the other hand is designed to correct a deficient situation. Problem-solving requires a root cause analysis so any corrections will prevent the problem from reoccurring. For example, if a valve was installed in a pipeline several months ago and now begins to leak, that is a problem. The problem-solver would determine the cause of the leak such as faulty installation, incompatible materials (gaskets, etc.) or incorrect valve. If the valve was replaced without finding the root cause, this valve may no longer pose a problem, but a similar problem could occur in some other location. If the gasket material was incompatible, the maintenance person repairing the valve may put the correct gasket in so that it does not leak any more. But if the cause was improper installation by the last maintenance person, or incorrect material purchased by the buyer, or a change in fluid in the pipe, a similar mistake could easily happen later on.

Understanding how to use JMT is important in problem-solving because it develops a questioning attitude so that people continuously ask 'why?' so they can get to the root cause. JMT is credited to be the origin of the "Five Whys" used in problem-solving. Also, if the problem-solver gets to the point in his analysis where a process is at fault, JMT should be used to correct that process.

and production positions. As a result, improvements will be made, not only on the shop floor, but also in purchasing, accounting, engineering, sales, human resources (HR), and even with senior management. JMT is a process-improvement program as opposed to a problem-solving program. However, as a process improvement program, it is a necessary, fundamental part of any problem-solving program, since many problems are centered on processes. Refer to Box 4.1 for additional information on problem-solving versus process improvement.

The process of JMT creates other benefits that occur as "by-products." Because the implementers must speak with others about their ideas, communication improves. As a result, teamwork improves, since people will know more about others' jobs and will be working together. Morale improves because people feel they have some autonomy, which they did not have before. All these benefits occur because employees feel they are part of the "team." "The company" has recognized their ideas and used them, and they were part of the effort. After one JMT 10-hour program, a woman told me that she had been at the company for 14 years, and this was the first time management asked for her ideas. That may or may not have been true, but her perception is her reality. She now felt more a part of the team.

Perhaps the greatest advantage of JMT can be seen with "The Hummingbird Effect,"[1] which states that one innovation can create other unanticipated and seemingly unrelated innovations or changes. The first of the six innovations that Johnson talks about in his book is glass. Gutenberg invented the printing press, which, of course, made books affordable to many more people. Literacy increased, and so did the number of people who would read books on a regular basis. This led to the discovery that most people were farsighted. They could see clearly at a distance but had trouble seeing objects that were close. Before the common use of books, people had no need to see small details. Spectacles had been invented, but the only people who had a need for them were those who had occasion to read on a regular basis, which would mainly be monks. As a result of the increased number of books, spectacles became more popular because now there was a need for them. With the demand for more spectacles came the demand for more lenses, which in turn increased the lens business. That meant more people were making lenses, and thus there were more people (everyone in the supply chain) in contact with lenses, since they were more common. That led to more research of lenses, which led to the invention of the microscope and the telescope. The microscope allows us to see very small objects. As someone was looking at a piece of cork, he saw that it was actually composed of cells, and was not really one solid piece. The telescope allows us to see objects at great distances, and this led Galileo to discover the first four moons of Jupiter. Gutenberg did not think that, by inventing a printing press, Galileo would discover moons orbiting Jupiter, but it led to it just the same. I grant that the invention of the printing press is a world-changing invention, but smaller inventions and innovations can also have an impact. A small improvement in one area of a facility can be seen by someone in another area and trigger a thought that will lead to unanticipated changes elsewhere. The more improvements that are made, the more likely this will happen.

JMT also enables another innovation concept called "adjacent possible."[2] Briefly stated, adjacent possible means that innovations are possible only when they are adjacent to innovations that already exist. Everyone has ideas, but some of those ideas cannot be fulfilled because the technology required to implement the idea may not have been developed yet. People have been thinking about flying for many years, and vertical flight, such as is done with a helicopter, can be traced back to 400 BC in China.[3] Leonardo da Vinci made a design in the 1480s, and Thomas Edison was given a grant in 1885 to make a helicopter. His calculations showed that a successful design would require an engine that weighed only 3 or 4 pounds per horsepower. Because the technology for such an engine did not exist in 1885, a helicopter was not flown successfully until the twentieth century. The implementation of ideas enable more ideas to be implemented. JMT is a good way to harvest as many ideas as possible, which will in turn stimulate other ideas.

Misconceptions of JMT

The implementation of JMT can spawn some misconceptions, mainly because it requires that people think about their jobs differently, and thus will actually change the culture of an organization. In the early 1940s, when it was first introduced, both engineers and workers did not like it. The engineers thought it would take away their jobs and the workers thought that they were doing extra work for which they were not paid. These same objections may be made today, although they may not be stated so plainly. In response to these concerns, the engineers were reassured first that there would always be enough work for improvements, and, second, that they deal with a wider span of processes and on several levels. The improvements implemented by operators will only be within the realm of the operators' work areas, while those of the engineers will cover a broader area. The operators know the fine details of the job better than the engineers, and so they are better able to see a need for certain improvements. Engineers are assigned a larger area to oversee, but both types of projects should enhance each other and not conflict. The reasoning with the operators is that it is everybody's job to improve whatever it is they do and the improvements they make will make their work easier. Furthermore, if both groups use JMT, they will be speaking with each other as they are working on their own projects.

A main principle of JMT is to question everything (see Principle 5, page 108.) When people's work is questioned, they often can become resistant. When a person is using JMT and s/he questions why something is done, the reason for the question is that the inquirer does not know the answer. S/he is not questioning the value or capability of the person who did the work, but is only seeking to understand it. Children 2–3 years old to about 8–10 years old are well known for asking "Why?". They are not questioning the validity of how something was done but only seek to know why it is. Children ask the question "Why?" repeatedly and the conversation usually stops when we do not know the answer. The person using JMT may not be so inquisitive, but the concept is the same. Thus, the person using JMT and asking the questions must be aware of the reaction of the person being asked and be suitably tactful.

Another misconception in the culture change will be that operators will now be spending some time making improvements. Thus, their work hours must be adjusted so this can happen. Supervisors will say that there is not enough time for this to happen, but the truth is that there is not enough time for it NOT to happen. Time can be found for having everyone involved in making improvements if the method is implemented slowly. There are various schemes for allocating time for improvements, but the simplest is to set aside 15 minutes a day, or an hour a week, devoted to improvements. The accumulation of improvements should result in having more time available and the initial time can be expanded as necessary.

These misconceptions should not be considered detrimental, but rather they should be seen as an advantage for management, engineers, and operators. Operators will have more input into what they do because they will have a formal, positive conduit for transmitting their ideas. Engineers will have assistance in projects because operators will be more involved and the engineers will be working on projects of a wider scope. Because more people will be involved in improvements, management will see a higher percentage of everyone's abilities, with a resulting rise in productivity.

Continual Improvement Programs and Suggestion Programs

Having a continual improvement (CI) program is currently a very popular idea in industry. Because of Lean thinking, people recognize now, more than ever before, that every company must continually improve all facets of the productivity of its operation. If the company does not, its competition will gain. Doing nothing is the same as moving backwards. Most CI programs, however, use a small group or department to achieve the improvements. Often, to these people's credit, they attempt to involve all employees, but because it is a CI department, people often think of it having a separate purpose. I'll do my job and they'll do their job, which is to make improvements. As mentioned, everyone has ideas on improvements to be made and the people closest to the work, those actually doing it, usually have the best, most effective ideas. The problem lies in the fact that most people do not know how to vet, sell, and implement their ideas. Vetting an idea is important because it must be determined to add value before it is used. Selling it, of course, is important because the best idea is useless if no one agrees to use it. Implementing is important; often the improvements are so small that a CI department might never get around to actually acting on the suggestion. These are the gaps that JMT fills so that the CI effort can truly be a company-wide program.

Continuous improvement through employee involvement is compared and contrasted with the suggestion system in a series of books about a method called Kaizen Teian.[4] These and other sources fully discuss the suggestion program used in many companies. A suggestion program is one that offers suggestion forms to be completed by anyone who has an idea for any improvement. The employee is to complete the form detailing the suggestion, and then deposit it into a suggestion box. The suggestions are collected periodically and evaluated by a committee or an individual, who usually is an engineer. The evaluator may contact anyone necessary if he needs additional information. He may also go to the person who originated the suggestion for additional details. The evaluator first determines if the suggestion will work, and then how much will be saved on a periodic basis. Based on the amount of savings, the employee who made the suggestion will be given an award. If the savings are positive but small, the suggestion will not be accepted.

The faults with such a program are many, but briefly they include

- Small improvements with little payback are not acted on—discouraging employees from making suggestions.
- Ignoring small improvements prevents larger improvements from surfacing.
- Ignoring small improvements eliminates long-term gains, which can be substantial.

- Large rewards will be given for suggestions that return significant savings—get employees to view it as a lottery: some ignore it completely and others "stuff" the suggestion box with extraneous thoughts.
- Significant time between when the suggestion goes into the box and when the employee receives a reply—discourages use.
- Employees have little involvement in the process aside from making the initial suggestion.
- Monetary rewards are counterproductive.[5]

The overall disadvantage of the suggestion program is that it is counterproductive. It actually discourages people from making suggestions. Operators have a good view of their own job or line or department, but they are not in a position to see the broader scope that would result in large paybacks. Operators also know their job better than anyone else, except for another operator. They can thus offer suggestions about what they do, but the gains will not be great since the scope of their job is limited.

In comparing the two programs, one might say that the suggestion program goes deep, looking for the largest payback. The Kaizen Teian program goes wide, looking for the most improvements. The philosophy behind the latter is that making many small improvements has several advantages over making a few large improvements. A main advantage is that suggestions beget suggestions. If my suggestion is accepted, as small as it is, I am encouraged to make another. Also, making ten suggestions that save $10 each is easier than making one suggestion that saves $100. That means the company should do whatever it can to encourage people to make suggestions. Accepting many small suggestions with little payback will increase the probability of getting suggestions with very large payback. This also develops an improvement culture where the employees drive the program. What Kaizen Teian does not discuss is how all employees learn how to vet, sell, and implement their ideas. Training such as JMT is the missing link.

Principles of Job Methods Training

The principles on which JMT is based follow the method itself, but there are some fundamental underlying ideas that should be emphasized.

1. Everyone Has Ideas

If you show someone a job that they have never seen before, they will, almost immediately, have questions about it. Shortly after that they will have formulated ideas on how to do some part of the job differently. They believe their ideas will make the job go better or faster. If that is true, why do employees not participate in company suggestion programs? Several reasons are listed in the section about suggestion programs, but the point here is that just because people do not participate in suggestion programs does not indicate that they do not have ideas about improving their jobs. The fact that everyone has ideas and is willing to share them is discussed thoroughly in *Ideas Are Free*,[6] written by Alan Robinson and Dean Schroeder. This principle is important because it makes us remember that JMT is for everyone and not a select few.

2. Most People Must Be Trained to Use Their Ideas

Those who recognize that everyone has ideas often begin to create a program to harness the collective creativity of the personnel in an entire organization. In creating such a CI program, they often assume all personnel know how to use their ideas correctly. If any training is done for this effort, it is usually limited to the core team who will be driving the effort. Although they often strive to get others involved in making the improvements, they may or may not transmit such training to the main population. If everyone in an organization is not trained properly in the CI effort, it will have limited success because everyone will not truly be involved.

All improvements start with an idea, but the first requirement is to vet that idea to determine whether or not it will add value to the process. If the idea does not add value, it should not be implemented. Once it has been decided that the idea will add value, it must be sold to everyone who has some influence in the selected area. If people who have a stake in the process that is affected by this idea do not agree with it, one of two actions will happen. Either the idea will die a sudden death or, what would be worse, the idea will be misused, fail to take hold, and die a slow, agonizing death. Once the idea has been vetted and sold, it must be implemented. When an idea brings significant savings, many people are aware of it and implementation is not a problem. However, if the payback is small and the idea is very local, a central implementation group may take extensive time to make the change, if they do it at all. Furthermore, even if the originator gets recognition for the idea, he may not be involved to any other degree.

I believe many people in charge of CI programs think everyone has the skills to vet, sell, and implement their ideas. The people in these positions are managers and/or engineers and they are in their position because they have these skills. Because they have had these fundamental skills for a long time, they often take them for granted and assume that everyone has them. I can empathize with these people who are managers and engineers, or "engineer types," because I too was one of them. Engineers are not only trained to vet, sell, and implement changes, but that is their main objective. As an engineer myself, I had assumed that everyone could do this at least to some extent. It was only within the last decade or so that I recognized that everyone does not think as I do (which is a good thing), but everyone can easily be taught to vet, sell, and implement his or her own ideas.

3. People Like to Share Their Ideas

Generally, people like to share their ideas when they are recognized for them. Although this may seem counterintuitive, recognition rather than a monetary reward is what most people seek. The sharing of ideas is affected by peer pressure, so that when one or two people in a group have been recognized for what they know or have done, others seek the same recognition. They will act on their ideas if they know how.

4. Break a Task into Small Details

When we do get an idea for improvement for something we do, that idea focuses on one aspect, set of actions, or perhaps, just one action of that job. Having ideas come to us "out of the blue," so to speak, is good and we should act on them, but we cannot depend on this method for improving our work in a continual manner. In order to have a method for improving our work in a continual manner, we must analyze the work using a repeatable approach. In doing this, we must break the task down into small details. The smaller we make the details, the better our improved method will be.

5. Question Every Detail: Question Everything

In the analysis noted above, we must also question every detail because no action should be "sacred." In order for any task to be as efficient as possible, each action must have a reason. If there is no reason for the detail, we should not be doing it.

6. Use the Six Questions, but in the Specific Order

Eliminate first; combine and rearrange next; improve/simplify last

Improvements happen when we ask the six questions familiar to all of us. However, the first question to ask is, "Why are we doing this?" That is the only question that leads to eliminating the detail. Eliminating unnecessary details is the easiest and fastest way to make improvements because once we eliminate the detail, we do not have to question it anymore. The reason it is not done more often is because some unnecessary details are difficult to see. Analyzing the task and breaking apart the details brings unnecessary details to light. Note that a follow-up question to "Why is the detail necessary?" is, "What is the purpose of the detail?" Although the two questions seem similar, finding the purpose gives us added information. For example, a reason for some details is that they follow standard policy. In order to know if that detail is necessary, we must know what purpose it serves in that policy.

7. Speak with Others

A job improvement should never be made in isolation, simply because one person does not know everything. We must always get information and opinions from others, even if we are the sole person doing the job. Someone else will ultimately touch whatever it is we do and so we must know the effect our job has on others.

8. Quantify

Whenever we get an idea we think will improve our operation, whether it is randomly or by using JMT, we stay with the idea when we become convinced it will make the intended improvement. However, at this point it is just an idea, and ideas can be misleading. Something can sound beneficial, but when we actually start doing it, we find out differently. As a result, the benefits of every change must be quantified. That is, if we believe a change will make an action faster, we must know how much faster. Improvements in some tasks can be difficult to quantify, yet we may find that when we break the task down into smaller details, the details themselves are easier to quantify. Putting numbers on improvements will verify to us that the change should be made, and it will also help us to convince others.

9. Get Approvals

Principle 7 says to speak with others. Not only will we get ideas from them, but often it is necessary to get their approval. In dealing with changes, people like to know what is going to happen before it does. If a person has the authority to block or accept your improvement, you should obtain that person's favor as quickly as possible.

10. Apply the New Method

The "death" of many ideas occurs only because they are never implemented. If your idea appears to be too ambitious, and you cannot implement the entire idea at this time, take a closer look and see if any part of it can be acted on. This will offer some benefit at least, and may also demonstrate that the remainder should be implemented.

Using Job Methods Training

As with the other two J programs, JMT should be used as a tool and not just as an exercise. It may be wasteful to have someone use JMT for a job that is not causing anyone any difficulty, because there probably is a job that does need this attention. Asking someone to "see if you can make some improvements to that job" without having a valid reason may well result in a time reduction, for example. But if that reduction is not reflected in an overall benefit, then the time spent was wasted. Sometimes a job just does not "feel" right and you know there has to be something that will make the overall flow smoother or faster. JMT can be applied here.

The example we will follow is a viscosity check shown in Figure 4.1. Refer to Figure 4.2 for the JMT pocket card. The scenario in this example is that an inspector must check the viscosity (thickness) of the contents of two tanks every 4 hours. There are two 5,000-gallon tanks involved and each one is 10 feet in diameter and 10 feet tall. Both are filled to a depth of about 8 feet. A mixer in each tank keeps the product mixed up. Stairs lead to a catwalk that surrounds each tank for access to the product. A conveyor operates above the tanks and a robot takes parts from the conveyor, dips them into each tank and places them back onto the conveyor. The inspector first turns off the power to the conveyor and the robot for safety reasons, so that he will not be harmed when he enters the robot area. He then gets his

Understanding JMT ■ 111

Job Methods Breakdown Sheet

Operation: __Viscosity check__ Product: _____ Molds Department: _____ Inspection

Originator(s): __Ernie south__ Assisted by: __Bob Wales, Chris Spear, Rob Wine, Karl Black__ Date: __5/20/15__

#	List all details {Present ~~Proposed~~} List every single thing that is Done	Distance (feet)	Time (minutes)	Notes Reminders, Tolerances, Etc.	Why? What?	Where? When? Who?	How?	Ideas Write them down Don't trust your memory
1	Walk to power switch	10						
2	Turn off conveyor power			To enter conveyor area				
3	Walk to tool cabinet	20						
4	Pick up zahn cup, stopwatch and screen							
5	Walk to white tank	30						
6	Take viscosity reading							
7	Record viscosity							
8	Walk to sink	25		To clean zahn cup & screen				
9	Clean zahn cup			For accurate reading				
10	Walk to blue tank	35						
11	Take viscosity reading							
12	Record viscosity							
13	Walk to sink	35						
14	Clean zahn cup			For accurate reading				
15	Walk to deionized water supply	50						
16	Get deionized water							
17	Walk to white tank	50						
18	Add deionized water			To correct viscosity				
19	Walk to blue tank	10						
20	Add deionized water			To correct viscosity				
21	Wait 15 minutes							
22	Return deionized water container	100						
23	Get zahn cup and stopwatch							
24	Walk to white tank	10		If NG, must get deionized water.				
25	Take viscosity reading							
26	Record viscosity							
27	Walk to sink	25						

Figure 4.1 MBS viscosity check present method (Step 1).

#	List all details { Present ~~Proposed~~ } List every single thing that is done	Distance (feet)	Time (minutes)	Notes Reminders, Tolerances, Etc.	Why? What?	Where? When? Who?	How?	Ideas Write them down Don't trust your memory
28	Clean zahn cup			For accurate reading				
29	Walk to blue tank	35						
30	Take viscosity reading			If NG, must get deionized water				
31	Record viscosity							
32	Walk to sink	35						
33	Clean zahn cup			For accurate reading				
34	Walk to power switch	55						
35	Turn on conveyor power			Resume production				
36	Walk to tool cabinet	20						
37	Replace zahn cup, stopwatch and screen							
	Totals	545	47					

Figure 4.1 (Continued) MBS viscosity check present method (Step 1).

Understanding JMT ■ 113

How to Improve
Job Methods

A practical plan to help you produce **greater quantities of quality products in less time** by making the best use of **Manpower, Machines, and Materials**, now available.

STEP I - BREAK DOWN the job.
1. List all the details of the job exactly as done by the Present Method.
2. Be sure details include all:
   - Material Handling
   - Machine Work
   - Hand work

Write details as you observe the job, NOT as you remember it.

STEP II - QUESTION every detail.
1. Use these types of questions:
   WHY is it necessary?
   WHAT is the purpose?
   WHERE should it be done?
   WHEN should it be done?
   WHO is best qualified to do it?
   HOW should it be done "in the best way?

2. Question everything, include the:
Materials, Machines, Equipment, Tools, Movement, Safety, Product Design, Layout, Workplace, Housekeeping

STEP III - DEVELOP the new method
1. **ELIMINATE** unnecessary details
2. **COMBINE** details when practical
3. **REARRANGE** for better sequence
4. **SIMPLIFY** all necessary details —
   - Make the work easier and safer
   - Pre-position materials, tools and equipment at the best places in the proper work area
   - Use gravity feed hoppers and drop delivery chutes
   - Let both hands and feet do useful work
   - Use jigs and fixtures instead of hands for holding work
5. Work out your ideas with others
6. Write up your proposed new method

STEP IV - APPLY the new method
1. Sell your proposal to the boss.
2. Sell the new method to the operators.
3. Get final approval of all concerned on Safety, Quality, Quantity, Cost
4. Put the new method to work. Use it until a better way is developed.
5. Give credit where credit is due.

Continually Improve the Method
JOB METHODS TRAINING PROGRAM
TWI Learning Partnership
www.TWIPartners.com
(585) 305-6820

Figure 4.2 JMT pocket card.

tools from a cabinet and checks the viscosity of the first tank. He cleans the tools and checks the viscosity in the second tank. If either tank is too viscous (thick), he adds some deionized water and rechecks the viscosity. The fluid will never be too thin since water will only evaporate (to thicken the mixture) and will not condense (to dilute the mixture). He then records the data on an inspection report, turns on the power, and returns the tools to the cabinet.

Describing How to Use JMT

In describing how to use JMT we will follow a four-step method.

Step 1: Break Down the Job

In JIT we broke down the job into macro advancing steps. In JMT, however, we must break down the job into much finer detail. The more finely we see the details, the more we can question, and the more we can find what does not add value. The tool we use in JMT is the job methods breakdown sheet or MBS. As with the job breakdown sheet (JBS) used in JIT, we will obtain more information if we watch someone perform the job, as opposed to us doing it or recalling it from memory. Although we will be recording more details in an MBS than we do steps in a JBS, it will be easier because we will record just what we see.

The first part of the MBS to complete is the heading, which is self-explanatory. The people in the "assisted by" field are those who helped the inspector in this analysis. It is always important to include the people who assisted you to make sure that everyone gets credit. We will find out later that the people include the department supervisor, an assistant (whom he watched do the job), a maintenance person, and a person from the safety committee. Notice that we will be using the same form for the "present" or existing method and for the "proposed" method. Additionally, notice that in the second box of the first row both "present" and "proposed" are listed. Since the first iteration will be for the "present" method, we will cross out "proposed." The MBS has nine columns. The first numbers the details. The second lists the details. Columns 3 and 4 are for recording distance and time so we can compare present with proposed quantitatively. Column 5 is for any notes that we think may be relevant or helpful later on. Columns 6–8 will help us remember to question every detail and give us the method for questioning every detail. Column 9 is for new ideas we get as we watch the job. As we perform the task, we will be scrutinizing everything we do, and more likely than not we will get some epiphanies or flashes of insight on how we can make the job better. Column 9 is for recording these epiphanies, flashes of insight, and ideas as they occur. *It is important to remember not to take any action at this time but rather to write down your ideas.* When people are reviewing a task intending to improve it and they act on the first idea they get, they often get sidetracked and do not finish the analysis. Also, an idea we get now might not even be applicable later if we decide to eliminate some steps. Let's review the present method and see how Ernie the inspector filled out the MBS.

Since he must first turn off the conveyor and the robot, he walks to the power switch. Notice that he did not write the first detail as "turn off conveyor" because he wants to be as detailed as possible. If a detail is not included, it cannot be questioned. Ernie records the distance walked but not the time of each detail. He must think that it is sufficient to have only the total time recorded. Notice also in Detail 6—take viscosity reading, he does not break down this operation into greater detail. By not finding smaller details, this analysis will not allow him to improve the actual checking procedure. He might think that, if necessary, he would do that later. Some judgments must be made about how much detail to create because time is always a factor. The same thinking could be used for "record viscosity" and "clean Zahn cup." As the details are written, Ernie has written some notes as they occur to him. This document can be kept on file for others to see when someone wants to review the viscosity check for possible improvements. The notes recorded will tell that person what Ernie was thinking and why he made the decisions he did. For example, someone may ask why it is necessary to clean the Zahn cup between each reading because time could be saved if that were eliminated. We can see that cleaning the cup is necessary for accurate readings. Ernie summed the distance and time and we see that he walked 545 feet and the job took about 47 minutes. Note that we are not intending to do a time study here but just want to know roughly how far the person has to walk, and how long it takes. The MBS can therefore provide a useful vehicle for creating a robust continual improvement system.

Step 2: Question Every Detail

We now take the MBS and question each of the details we have written. (Refer to Figure 4.3.) We ask the six basic questions: Why? What? Where? When? Who? How? They must be asked in that order. "Why" and "What" lead

116 ■ *The TWI Facilitator's Guide: How to Use the TWI Programs Successfully*

Job Methods Breakdown Sheet

Operation: Viscosity check   Product: Molds   Department: _____   Inspection: _____

Originator(s): Ernie south   Assisted by: Bob Wales, Chris Spear, Rob Wine, Karl Black   Date: 5/20/15

#	List all details {Present / ~~Proposed~~} List every single thing that is done	Distance (feet)	Time (minutes)	Notes Reminders, Tolerances, Etc.	Why? What?	Where? When? Who?	How?	Ideas Write them down don't trust your memory
1	Walk to power switch	10						
2	Turn off conveyor power			To enter conveyor area				
3	Walk to tool cabinet	20				✓		Move tools closer to slurry tank
4	Pick up zahn cup, stopwatch and screen							
5	Walk to white tank	30				✓		Move tools closer to slurry tank
6	Take viscosity reading							
7	Record viscosity							
8	Walk to sink	25		To clean zahn cup and screen		✓		Water supply nearer tanks
9	Clean zahn cup			For accurate reading			✓	Better way?
10	Walk to blue tank	35						
11	Take viscosity reading							
12	Record viscosity							
13	Walk to sink	35				✓		Water supply nearer tanks
14	Clean zahn cup			For accurate reading			✓	Better way?
15	Walk to deionized water supply	50				✓		Deionized water nearer tanks
16	Get deionized water							
17	Walk to white tank	50				✓		Deionized water nearer tanks
18	Add deionized water			To correct viscosity				
19	Walk to blue tank	10						
20	Add deionized water			To correct viscosity				
21	Wait 15 minutes							
22	Return deionized water container	100				✓		Deionized water nearer tanks
23	Get zahn cup and stopwatch							
24	Walk to white tank	10		If NG, must get deionized water.				
25	Take viscosity reading							
26	Record viscosity					✓		
27	Walk to sink	25						Water supply nearer tanks

Figure 4.3 MBS viscosity check present method (Step 2).

#	List all details {Present ~~Proposed~~} List every single thing that is done	Distance (feet)	Time (minutes)	Notes Reminders, Tolerances, Etc.	Why? What?	Where? When? Who?	How?	Ideas Write them down don't trust your memory
28	Clean zahn cup			For accurate reading			√	Better way?
29	Walk to blue tank	35						
30	Take viscosity reading			If NG, must get deionized water				
31	Record viscosity							
32	Walk to sink	35						Water supply nearer tanks
33	Clean zahn cup			For accurate reading		√		
34	Walk to power switch	55					√	Better way?
35	Turn on conveyor power			Resume production				
36	Walk to tool cabinet	20						
37	Replace zahn cup, stopwatch and screen							
	Totals	545	47					

Figure 4.3 (Continued) MBS viscosity check present method (Step 2).

to eliminating, and if a detail is eliminated, there is no need to go further with the questions. "Where," "When," "Who" lead to combining and rearranging. Combining and rearranging may affect the last question: "How." "How" leads to simplifying. When we do this it is much more effective if we focus on the verb, since that is the action word. The first detail is "Walk to power switch."
Analysis:

- Why walk? Power switch is away from tanks (on wall).
- What? Purpose is to turn off power.
- Where? Best done at this station since will minimize machines down.
- When? Right before viscosity check to minimize production stoppage.
- Who?[7] Inspector has been approved by maintenance and safety to control power switch.
- How? Easiest way.[8]

We repeat this procedure for each detail. Note Detail 3—"Walk to cabinet."
Analysis:

- Why walk? Because the tool cabinet is away from the tanks. Why?
- What? Purpose is to get tools.
- Where? Best place for tools is between the tanks since they are dedicated to the viscosity check and will not be used anywhere else.
  Add idea: "Move tools closer to tank." Put a check mark in Column 7, "Where? When? Who?"
- When? Get tools when needed. Store in cabinet so they don't get lost.
- Who? Inspector is person who uses them.
- How? Cabinet does not need modification.

The first time a person uses the MBS it is best to actually ask each question for each detail. It is not necessary to write an answer to each question, but it is important to get into the habit of asking each question individually. As one becomes more familiar with the method, this process can be shortened, but care must be taken not to revert to assumptions about the job. We are always looking for the best place, the best time, and the best person to do the detail. Remember that the best person to do the detail is not the best qualified person, but rather the person whose qualifications and abilities best match the requirements of the detail. As you scan Figure 4.3 you see that Ernie added some notes in Column 9 and did not take any action at this time. He recognized that he had to walk 20 feet to get the tools and did not know why he had to do that. Why are the tools not closer to the tanks? He must investigate to find out. He also put a check mark in the appropriate Column 6–8 to remind him of what led to his questions.

Step 3: Develop the New Method

Ernie must now see what improvements can be made to this job and again, he goes detail by detail. Any detail that did not result in a satisfactory answer to "Why?" or "What?" might result in it being eliminated. Ernie looks down Column 6 and does not see any check marks, which means all the "Why?" questions were answered satisfactorily. He then scans Column 7, "Where? When? Who?" and sees that Detail 3 has a check, and his note is to see if the tools can be moved closer to the tanks. This takes him to the maintenance department, where he asks if the tool cabinet can be mounted on a tank, or on the stairs going to the catwalk, or be freestanding near the tanks. Maintenance says they can easily move the tool cabinet and mount it on the stairs.

Notice that he also had to walk to the sink to clean his tools, and to the deionized water. He asked if they could be moved closer to the tanks. The deionized water was in a 55-gallon drum and it could be located anywhere, but the sink had a drain and it would be too expensive to bury pipes in the floor. However, Ernie did not use much water to clean the cups. He calculated that he used less than 50 gallons a month. He asked maintenance if he could have a drum of water and an empty drum with a sink mounted on it. Would maintenance be willing to empty the dirty water tank once a month? They said they would.

Based on this analysis, Ernie took a new MBS form and wrote his proposed method (Figure 4.4). Recall that we are using the same form for the "present" or existing method and for the "proposed" method. Additionally, notice that in the second box of the first row both "present" and "proposed" are listed. Since this iteration will be for the "proposed" method, we will cross out "present."[9] Because he could move the deionized water, the sink and water supply, and the tool cabinet between the tanks, he was able to remove two details—returning the deionized water, and walking (back) to the tool cabinet. However, that is all that could be eliminated. He did not eliminate returning the tools but combined it with walking to the water supply to clean the tools. The distance was reduced from 545 feet to 90 feet, and the time was reduced from 47 minutes to 31 minutes. Note that this

120 ■ *The TWI Facilitator's Guide: How to Use the TWI Programs Successfully*

Job Methods Breakdown Sheet

Operation: __Viscosity check__   Product: _____   Department: __Molds__   Inspection: _____

Originator(s): __Ernie South__   Assisted by: __Bob Wales, Chris Spear, Rob Wine, Karl Black__   Date: __5/27/15__

~~Present~~ / Proposed

#	List all details List every single thing that is done	Distance (feet)	Time (minutes)	Notes Reminders, Tolerances, Etc.	Why? What?	Where? When? Who?	How?	Ideas Write them down Don't trust your memory
1	Walk to power switch	10						
2	Turn off conveyor power			To enter conveyor area				
3	Walk to tool container	10		Moved between tanks				
4	Pick up zahn cup, stopwatch and screen							
5	Walk to white tank	5						
6	Take viscosity reading							
7	Record viscosity							
8	Walk to water supply	5		Water supply moved between tanks				
9	Clean zahn cup			For accurate reading				
10	Walk to blue tank	5						
11	Take viscosity reading							
12	Record viscosity							
13	Walk to water supply	5						
14	Clean zahn cup							
15	Walk to deionized water supply	5		Deionized water moved between tanks				
16	Get deionized water							
17	Walk to White tank	5						
18	Add deionized water			To correct viscosity				
19	Walk to Blue Tank	5						
20	Add deionized water			To correct viscosity				
21	Wait 15 minutes							
22	Get zahn cup and stopwatch							
23	Walk to white tank	10		If NG, must get deionized water.				
24	Take viscosity reading							
25	Record viscosity							
26	Walk to water supply	5						

Figure 4.4 MBS viscosity check proposed method (Step 3).

#	List all details ~~Present~~ Proposed List every single thing that is done	Distance (feet)	Time (minutes)	Notes Reminders, Tolerances, Etc.	Why? What?	Where? When? Who?	How?	Ideas Write them down don't trust your memory
28	Clean zahn cup			For accurate reading				
29	Walk to blue tank	5						
30	Take viscosity reading			If NG, must get deionized water				
31	Record viscosity							
32	Walk to water supply	5						
33	Clean zahn cup			For accurate reading				
34	Replace zahn cup, stopwatch and screen							
35	Walk to power switch	10						
36	Turn on conveyor power			Resume production				
	Totals	90	31					

Figure 4.4 (Continued) MBS viscosity check proposed method (Step 3).

check is performed twice a shift and there are two shifts per day, so the daily savings are 64 minutes per day and 1820 feet walked per day. This is not only a reduction in inspection time, but since production is down when the robot and conveyor are down, this added 64 minutes per day to available production. Now Ernie had to implement his proposal.

Step 4: Apply the New Method

Unless you are working completely by yourself, it is always important to convince anyone who has any influence with the method of the value of the changes you propose.

Therefore, the first part of applying the new method is to convince the person to whom you report. This should not usually be a problem because you should have been working with this person on the improvement. If nothing else, you should have been keeping him or her aware of your activities. Second, you must also sell the new method to the other operators who will be using it, and they also should have been involved in what you have been doing. The third activity in applying the method is to get any necessary approvals. The method has not been implemented yet, so now is the time to make sure the changes are agreeable with all areas of the organization. This would include safety, engineering, quality, maintenance, scheduling, and so forth. Convincing people to change a method and getting their approvals are not always easy tasks.

The Proposal Sheet

The tool we use to accomplish this is the Job Methods proposal sheet, which has four main sections. Refer to Figure 4.5.

Heading

The heading is self-explanatory, but it should be emphasized that the field "Assisted by" is important because people should receive the recognition they have earned. Misuse of this field or ignoring it completely can significantly hurt a continual improvement program.

Summary

The summary should be a brief description of the previous and present methods, and the main gains. The person reading the summary does not want to know all the details involved.

Results

This is the heart of the proposal sheet because it makes the most convincing argument and it differentiates the JMT program from standard suggestion programs. The first person to be sold on the improvement is the person driving it. Whether this data does or does not convince the person making the

Job Methods Proposal

Submittedto: JimAnderson	Date: 5/25/15
Created by: Ernie South	Dept./Cell:
Assisted by: B. Wales, C. Spear, R. Wine, K. Black	
Product: Auto line	
Operations include: Viscosity check	

Summary
By moving the viscosity instruments, a water supply and a deionized water supply closer to the slurry tanks, we were able to reduce checking time by 34%. This means that the conveyor can now be operating an additional 32 minutes a shift or 64 minutes a day. (The conveyor must be shut down while the viscosity check is being done.)

Results

Metric	Before	After	Difference
Quality			No change
Cost		$210	$210
Production	13 hrs/day	14 hrs/day	7.7% increase
Safety			
Machine use			7.7% increase
Reject rate			
Number of operators			
Distance	545 ft	90 ft	555 ft
Time	47 min.	31 min.	16 min. (34% reduction)

Content

Tool board for viscosity instruments	$50
Run water supply to tanks	$160
Reservoir of deionized water next to tanks	No charge

The water supply is run to a portable sink that rests on a 55 gal. drum. Estimating the usage to clean the zahn cup, we believe that the drum must be emptied about once a month. The maintenance department supervisor and techs said they could easily include that removal in their monthly clean-up.

The deionized water is stored in 55 gal. drums and the material handler said she could put one anywhere in the department.

Figure 4.5 JMT proposal viscosity check.

Attach present and proposed breakdown sheets, diagrams, and other related items as appropriate.

Before turning in your PROPOSAL, be SURE you have rechecked the New Method with the Job Methods 4-STEP PLAN.

STEP I – BREAK DOWN the job.
1. List ALL the details of the job EXACTLY as done by the present method
2. Be sure details include all:
    Materials handling
    Machine work
    Hand work

STEP II – QUESTION every detail. Use these types of questions:
1. WHY is it necessary?
2. WHAT is its purpose?
3. WHERE should it be done?
4. WHEN should it be done?
5. WHO is the best qualified to do it?
6. HOW should it be done "in the best way?

Also, QUESTION the -

MATERIALS
Can better, less expensive or less scarce materials be substituted?
Can the scrap from this job be used for another product?
Have the defects and scrap been reduced to a minimum?
Are the material specifications entirely clear and definite?

MACHINES
Is each operating at maximum capacity?
Is each in good operating condition?
Are they serviced regularly?
Is the machine best for this operation?
Should a special set-up man or the operator make all the set-ups?
Can use be made of the machine's or operator's "idle" time?

EQUIPMENT AND TOOLS
Are suitable equipment and tools available?
Have they been supplied to operators?
How about gauges, jogs and fixtures?
Have equipment, tools, been properly pre-positioned to permit effective work?

PRODUCT DESIGN
Could quality be improved by a change in design or specification?
Would a slight change in design save much time or materials?
Are tolerance and finish necessary?

LAYOUT
Is there a minimum of backtracking?
Are the number of handlings and the distances traveled at a minimum?
Is all available space being used?
Are aisles wide enough?

WORK-PLACE
Is everything in the proper work area?
Can gravity feed hoppers or drop delivery chutes be used?
Are both hands doing useful work?
Has all hand-holding been eliminated?

SAFETY
Is the method the safest as well as the easiest?
Does the operator understand all safety rules and precautions?
Has proper safety equipment been provided?
Remember, accidents cause WASTE of manpower, machines, and materials!

HOUSEKEEPING
Are working and storage areas clean and orderly?
Is "junk" taking up space that could be used for additional operators, machines, benches, and operations?
Do away with anything that is unnecessary.
Be sure necessary things are in proper places.
See that good Shop Housekeeping reduces delays, waste, and accidents!

STEP III – DEVELOP the new method
1. ELIMINATE unnecessary details.
2. COMBINE details when practical.
3. REARRANGE for better sequence.
4. SIMPLIFY all necessary details.

STEP IV – APPLY the new method
1. SELL your PROPOSAL to your boss and operators.
2. Get final approval of all concerned on SAFETY, QUALITY, QUANTITY, COST.
3. Put the new method to work. Use it until a betterway has been developed.
4. Give CREDIT where credit is due.

Figure 4.5 (Continued) JMT proposal viscosity check.

change, it will have the same effect on all others. Without this data, the person making the change may think s/he has a good idea, but the data proves the validity of the idea. The magnitude of the gains is unimportant as long as it is positive. What is necessary is that all facets of the situation have been considered and all necessary parties contacted. The concept being used here is to proactively take away reasons why a change should not be made.

A reviewer who looks at this should immediately see that for an outlay of $210, we will gain an hour of production time. That should not be a difficult decision for anyone to make since the $210 will be a one-time cost, while the 64 minutes will be saved every day. Note that this improvement will add 32 minutes to the production time per shift. Although that sounds positive to Ernie, he must check with production control and relevant supervisors to confirm that the impact will be positive. He may have some knowledge of other departments but not enough to determine if an increase of 32 minutes in conveyor run time is beneficial or harmful to others. Most importantly, he should not make an assumption.

Content

This is the area when one can supply additional details. Additional sheets of data or drawings can be supplied, but the proposal sheet should be kept to one page. Therefore, this "content" section should be limited to one-third to one-quarter of a page. The proposal sheet would then act as a cover page for the remaining data. People can review the additional data if they are so inclined, but the main arguments should be on the front page. Ernie explained the origin of the costs and basically what had to be done.

Finally, applying the new method means implementing it. This should be done as soon as possible and if the complete change cannot be made all at once, you should proceed with as much of the plan as possible. Once approved, an area between the tanks could be selected for the deionized water and the portable sink. The area could be laid out and both of these items could be placed there. The tool cabinet would take some more time to move, and running the water to the sink even more time. However, whatever can be done should be done as soon as possible. There is no need to wait for everything to be done at once. If only the deionized water can be moved in the week after approval, the 64 minutes per day will not be saved, but some lesser amount will be. The JMT approach makes it very clear that since Ernie initiated the improvements, he must follow up on them. Each organization will devise its own system for this, such as a project board placed in a main area. Such a public display board serves two purposes. It provides peer pressure to prevent improvements from stalling and it acts as a medium to transfer ideas. Note also that since the process has changed, the JBSs associated with them must also change. It may not be Ernie's responsibility to edit an existing JBS, but it could be his responsibility to make sure it gets done. His supervisor would help him with this to whatever extent is necessary.

Epilogue to the Viscosity Check Example: Making Changes

We are reminded on the pocket card to "Continually Improve the Method." Once Ernie took a detailed look at this job, he continued to think about it because he felt somehow he was missing something.[10] He kept on thinking about Detail 2—"Turn off Conveyor and Robot Power." He knew why he had to do that but wondered if he was missing something. He created a new MBS with added details (Figure 4.6). When he got to 5—"Walk to white tank", he added 6—"Climb stairs" and 7—"Enter catwalk." The six details were easily answered for Detail 6, but here is the analysis for Detail 7:

Analysis Detail 7

- Why enter? To access slurry; why there? Only place; why? Is it possible to get sample and add deionized water while not on catwalk? Better way?
- What is purpose? Get sample
- Where should it be done? On catwalk; Where on catwalk? Anywhere because robot is not running. Is there a safe place to get sample when robot is operating? Check with safety committee
- When enter catwalk? Check with safety committee
- Who is best qualified? Inspector
- How to do in the best way? May be better way

After this analysis, Ernie had two more ideas to check out. He saw maintenance to find out what it would take to put a flange and nozzle on each tank so that he could take a sample without climbing the stairs. The maintenance person thought that it would be pretty expensive, and it could not be done until next summer during their annual shutdown for cleaning when the tanks would be empty. In the meantime, maintenance would investigate some more and get a better estimate of cost. Ernie could not proceed with that idea for at least 7 months, so he considered his second idea. He went to the Health and

Job Methods Breakdown Sheet

Operation: Viscosity check Product: Molds Department: Inspection

Originator(s): Ernie south Assisted by: Bob Wales, Karl Black Date: 6/14/15

#	List all details { Present ~~Proposed~~ } List every single thing that is done	Distance (feet)	Time (minutes)	Notes Reminder, Tolerances, Etc.	Why? What?	Where? When? Who?	How?	Ideas Write them down don't trust your memory
1	Walk to power switch	10						
2	Turn off conveyor/robot power			To enter conveyor area				
3	Walk to tool cabinet	10						
4	Pick up zahn cup, stopwatch, and screen							
5	Walk to white tank	5						
6	Climb stairs							
7	Enter catwalk							
8	Take viscosity reading			If NG, must get deionized water.				
9	Record viscosity							
10	Walk to water supply	5		Water supply moved between tanks				
11	Clean zahn cup			For accurate reading				
12	Walk to blue tank	5						
13	Climb stairs							
14	Enter catwalk							
15	Take viscosity reading			If NG, must get deionized water.				
16	Record viscosity							
17	Walk to water supply	5						
18	Clean zahn cup			For accurate reading				
19	Walk to deionized water supply	5		Deionized water moved between tanks				
20	Get deionized water							
21	Walk to white tank	5						
22	Climb stairs							
23	Enter catwalk							

Figure 4.6 MBS viscosity check new present.

#	List all details {Present ~~Proposed~~} List every single thing that is done	Distance (feet)	Time (minutes)	Notes Reminders, Tolerances, Etc.	Why? What?	Where? When? Who?	How?	Ideas Write them down don't trust your memory
24	Add deionized water			To correct viscosity				
25	Walk to blue tank	5						
26	Climb stairs							
27	Enter catwalk							
28	Add deionized water			To correct viscosity				
29	Wait 15 minutes							
30	Get zahn cup and stopwatch							
31	Walk to white tank	10						
32	Climb stairs							
33	Enter catwalk							
34	Take viscosity reading			If NG, must get deionized water.				
35	Record viscosity							
36	Walk to water supply	5						
37	Clean zahn cup			For accurate reading				
38	Walk to blue tank	5						
39	Climb stairs							
40	Enter catwalk							
41	Take viscosity reading			If NG, must get deionized water.				
42	Record viscosity							
43	Walk to water supply	5						
44	Clean zahn cup			For accurate reading				
45	Replace zahn cup, stopwatch and screen							
46	Walk to power switch	10						
47	Turn on conveyor power			Resume production				

Figure 4.6 (Continued) MBS viscosity check new present.

Safety Director and asked if he could walk on the catwalk while the robot was running. The stairs were out of the conveyor and robot area, as was half of the catwalk. Would it be possible for him to take a reading without turning off the power if he stayed on the half away from the robot? The safety person said that Ernie could do that but he would want to put up some safety chains to prevent people from accessing the "active" half of the catwalk. If the viscosity checks are done without turning off the robot and the conveyor, production would increase by an additional hour per shift, since the viscosity check now takes 31 minutes and is done twice per shift. The robot would run all day and the viscosity checks would not shut it down.

An important point of this story is that, although Ernie had been doing this job for several years, he never questioned what he was doing. Once he was given a systematic method to make improvements, he began to think more deeply about the job, which resulted in improvements. Furthermore, although his first attempt was very credible and resulted in a significant gain of an hour of production time, he kept thinking about it and achieved an even larger gain of 2 hours of production time. By breaking a detail into smaller details, Ernie created more actions to question. The more actions we have to question, the finer our improvement will be. Obviously, such large gains will not always happen, but what does happen is that once people start questioning their work, gains are usually made because there is significant waste in much of what we do. Once you give people a tool to make improvements, results can be impressive. After Ernie's second attempt, it might seem that little more can be improved. Future gains will not be as great since the only savings now will be found in reducing the inspector's time. However, the 31 minutes it now takes can be reduced somehow. Perhaps Ernie will find a faster way to clean the Zahn cup or he could write a separate MBS for what is now a single detail of taking a viscosity reading.

Although Ernie's improvements were made over a few weeks' span, others might occur over a longer time. Once JMT has been ingrained in an organization for a while, it will accumulate a file of MBSs. When someone thinks of an improvement of a job, one of the first steps to be taken should be to check the file and see if there is an MBS on record. This will give the person an overview of what has been tried and why. When a person gets an idea for an improvement of an existing job and mentions it to a person who has been doing the job, a remark often heard is, "We tried that before and it didn't work." No reason is given for why it didn't work, and often the idea dies right there. A well-written MBS could supply the information that would make this attempt worthwhile.

Auditing

Auditing is not necessary with JMT as it is for JIT. There should be JMT coaches who would be available to assist employees with completing a proposal sheet, but often the supervisor should do that. JMT will help employees develop a questioning attitude, as it did for Ernie, and a person's supervisor (coach) can help with this. I also discourage the use of a quota system where everyone must submit an improvement every month. Thinking of an improvement is not like solving a puzzle because it involves some creativity. A person may have three ideas in one month and none for three months after that. If the quota is one idea per month, this person may hold back two ideas so that he meets the quota. As a result, the productivity gains from those two ideas will be lost for those two months. Forcing someone to submit an idea when they have none can lead to lessening the effect of the program. The desire to be recognized will stimulate some people to offer improvements once they have the JMT vehicle. Others will be drawn by peer pressure. Once this starts, it will grow.

Documentation

A documentation system is necessary for JMT even if there is no formal continual improvement system. Many organizations such as the International Organization for Standardization (ISO) require documentation for all improvements and changes, and the JMT proposal sheet facilitates this requirement. The proposal sheets are useful when someone wants to make an improvement in an area that has already been addressed with JMT. People's thoughts on what has and has not been tried will be more evident. This documentation should be kept in a central location so it is easily accessible.

Notes

1. Johnson, S. 2014. *How We Got to Now: Six Innovations That Made the Modern World*. New York: Penguin, p. 5.
2. Johnson, S. 2007. *Where Good Ideas Come from: The Natural History of Innovation*. New York: Penguin, p. 31.
3. Helicopter. 2015. https://en.wikipedia.org/wiki/Helicopter.

4. Japan Human Relations Association. 1991. *Kaizen Teian 1*. Portland, OR: Productivity Press; Japan Human Relations Association. 1995. *The Improvement Engine*. Portland, OR: Productivity Press; Kaizen means continual, gradual improvement and Teian means proposal or suggestion.
5. Kohn, A. 1993. *Punished by Rewards: The Trouble with Gold Stars, Incentive Plans, A's, Praise, and Other Bribes*. Boston: Houghton Mifflin.
6. Robinson, A.G. and Schroeder, D.M. 2004. *Ideas Are Free*. San Francisco, CA: Berrett-Koehler.
7. Note that when asked "who?", the answer is whoever is best suited for the job. That means the person whose skills most closely match the job requirements. We want to avoid having a skilled operator performing a function a novice could do well, since that would be a waste of the skilled operator's talents.
8. When this JMT was originally written, manually turning off a circuit breaker was the best way. However, with the advanced electronics of today, a remote control may be possible. This would have to be reviewed for cost savings and safety considerations.
9. Computer tablets are becoming more common in the workplace and they are a good tool to use to record a JBS or MBS since additions and/or corrections can be easily made. Most are still done with paper and pencil but the writing of so many details is not as tedious as it might seem. Also, in addition to just crossing out "present" or "proposed" based on which method is described, some people like to highlight or circle the method being described.
10. This example was taken from a 10-hour JMT session. Ernie was the first participant to use JMT and after he gave his example on Tuesday, he continued to think about it during the week. Had he been more familiar with JMT, he might have questioned Detail 2: "Turn off conveyor power" more deeply during his first analysis. This example shows how JMT helps us develop a questioning attitude.

Chapter 5

Understanding Job Relations Training

Before cars, make people.

Eiji Toyoda

Job Relations Training is a short, intensive program that uses foundational concepts and a four-step plan to teach skills both in sizing up situations and in understanding and working with people.

Objectives of JRT

The objective of Job Relations Training (JRT) is to build strong, positive personal relationships among all members of an organization in order to prevent any personnel situations from interfering with production. If such detrimental personnel situations do occur, however, a secondary objective is to identify and correct them as soon as possible using a standard method. Thus, the main objective of JRT is to prevent personnel problems from interfering with production. No one is naïve enough to believe that any organization, no matter the size, will never have a personnel problem, so the secondary objective is to solve it as soon as possible using a standard four-step method. Each supervisor is to use a list of basic personnel foundations, which are stated as principles, to prevent problems from occurring. There is no standard method for using these principles. Each supervisor should keep them in mind and use them as each situation warrants. In the 10-hour training program, whatever foundations are used in a particular case study will be discussed, and any not used for a given case study will not be discussed. The four-step method, however, will be used and discussed for each case study during the week. The four-step method has a tendency to overshadow the foundations when delivering the JRT program. A mistake often made in delivering the JRT is to focus on problem-solving instead of problem prevention. This happens because while there is a standard four-step plan for problem solving, there is no standard plan for problem prevention. Therefore the emphasis in the 10-hour program should be on the foundations for problem prevention rather than on the four-step method for problem-solving. Both are important, but the Service emphasizes, "a campfire is easier to extinguish than a forest fire."

Benefits of JRT

The main and overall benefit of using the JRT method is that employees will work better together because they will have strong, positive personal relationships with each other. This is accomplished through three main efforts:

1. Using the foundational principles
2. Using the four-step method
3. Verifying all supervisors know how to handle personnel problems correctly

First, using the foundations results in having people treated as they would want to be treated. This leads to improvements in teamwork and morale. Second, when a negative personnel situation does arise, the supervisor who addresses it does so with a standard method that every other supervisor uses. The use of a standard method is important for at least two reasons. First, all supervisors will be using the same method and thus everyone will be given the same attention when situations arise. The result of this is that all employees recognize that they are being treated fairly. In addition, many supervisors may recognize a problem occurring but, without a standard method, they may not know how to address it. When the problem is large and they "run into it," they will do their best to resolve it because they are forced to. When a problem is small, however, they can easily say it will go away by itself if they have no standard way of resolving it. Once all supervisors have learned to use JRT well, large problems will be less likely to happen since they will be resolved when they are smaller and more manageable. Finally, people will enter the supervisory ranks who otherwise might not. There are instances where a person knows his job well and may be well liked by those close to him, and thus is given supervisory responsibility. Sooner or later, it is discovered through some incidents that he does not know how to deal with some personnel situations. Not much good can come of this. Bad feelings can arise if he stays in the position and bad feelings can also arise if he is demoted. Good potential supervisors may be lost because they do not intuitively know the skills of building strong, positive relationships or dealing with personnel problems.

Principles of Job Relations Training

The concepts promoted by the TWI Service were very far ahead of their time. For example, they valued diversity and having respect for people. When few women were in the production workplace, they stated repeatedly that the only difference between a man and a woman in the workplace should be what restroom they used.[1] Because the people of TWI had ideas that we consider to be contemporary, and thus have stood the test of time, many of those ideas stated in JRT could be considered to be principles. As such, we ignore them at our peril because each idea stated adds to the overall success of using JRT. A main tenet of Lean and an underlying facet of the Toyota Production System is that everyone should show respect, regard, or consideration for every other person. No one should be ignored. Because of that, JRT could be considered to be the most important of the J programs in that it is the linchpin that holds an organization together. The basis of that thinking lies in adhering to many of the following principles:

1. The strengths and successes of an organization are based on the strong, positive personal relationships of the people in it.
    An organization, by definition, is composed of a group of people. Every individual in an organization has a relationship with at least one other person in that organization. The stronger and more positive those relationships are, the more successful the people, and thus the organization, will be.
2. Treat everyone as an individual.
    This means that everyone should be treated fairly but not necessarily the same way. Coach John Wooden was the UCLA basketball coach from 1948 to 1975. His teams won 10 NCAA Division 1 Tournament

Championships from 1964 to when he retired – in 1975. A major reason for his phenomenal success was that he knew how to handle people. As people studied his techniques, some recognized that he used the principles espoused by the TWI J programs. One of those principles is this one, about which he said, "I believe, in order to be fair to all students, a teacher must give each individual student the treatment he earns and deserves. The most unfair thing to do is to treat all of them the same."[2] It is interesting to note that, although he was a coach, John Wooden referred to himself as a teacher.

3. Let each person know how s/he is doing.

   The annual review is one of a supervisor's duties that is often abused because it often comes as a surprise to the person being reviewed. The person is not surprised that s/he is being reviewed or even when it will occur. The surprise is the content of the annual review. The annual review should be a summary of the year's performance and both the supervisor and the direct report should be fully aware of all of that performance. The supervisor should review both what the person has done right and what s/he has done wrong as soon as possible after it happens and not just during the annual review. Feedback should be as immediate as possible.

   Supervisors often find it easier to give praise for something done right than correction for something done wrong, but one is as important as the other. As a result, some supervisors neglect to give negative feedback when something is done wrong. They may not be comfortable engaging in such a conversation and convince themselves that it is not necessary to tell the person about it. A principle of TWI that covers all J programs is that people want to be productive and do the right thing. People do not intentionally do the wrong thing. What can happen is that a person acts in a way with which his supervisor disagrees, but does not receive feedback. The person assumes nothing is wrong, so the same action may reoccur. When the supervisor does give feedback, it is, most likely, in the form of a reprimand because now the situation is larger than the first occurrence. This may be confusing to the person and degrade the relationship between the two. If the supervisor believes that the person wants to do what is right, then it is easy to turn this "negative" situation into a "positive" one by creating a "teaching moment." If the person knows he made a mistake, the supervisor and the person together should discover why the mistake was made and how to prevent it from reoccurring. If the person believed s/he was acting correctly, s/he lacks some instruction, which should be furnished by the supervisor or some other appropriate instructor. Most people can do most jobs but there are some jobs for which some people are not suited. People should not be capriciously removed from a job because it is thought that it does not "suit" them. Every attempt should be made to see that the person has the aptitude, attitude, and prerequisite knowledge for that job.

As organizations become flatter, supervisors will have more direct reports and will be unable to give each one the attention deserved. The number of direct reports a supervisor can properly oversee will depend on many factors, including the type of job and the quality of the training each person has received. This factor must be taken into consideration when evaluating a personnel situation in the organization.

4. Give credit when due.

Situations arise when a person contributes more than what is normally expected. When that effort is recognized it is reinforced and will be more likely to reoccur. This results in both the person and the organization growing. In order for the recognition to have any effect, it must occur as close to the contribution as possible. When recognition for extra effort is not given, that extra effort will become more infrequent. Some organizations have small rewards available to supervisors to give to employees for better than requested service. Some supervisors limit their praise to actions that warrant these rewards, but that neglects much action where the employee should be given credit, but that might not require a quantified award. A simple, "Thanks very much. I appreciate you doing that" can go far in building better relationships.

5. Tell people about changes that will affect them.

One of a person's basic needs is autonomy, which means having some control over what one does. This does not mean that everyone has to be 100% in control of everything that affects them. Not everyone wants that much control and it is unrealistic to expect it. However, people do like to know some of what the future holds so that they can mentally and perhaps physically be prepared for it. This principle is similar to the one above because it gives the person some autonomy and time to think about how to react to the change. There may be some instances when it is unwise to give people time to consider the change; but in using the JRT method we should consider all possible results of all of our actions.

6. Make the best use of each person's abilities.

   This is sometimes cited as the eighth waste in Lean thinking. Cross training is important because most organizations cannot afford to have one person perform just one job. However, it is equally important for everyone to perform a job for which they are suited. What this means is that the supervisor should know everyone's abilities so that the person can be used most effectively. The TWI Service believed that training "can be suitable to the individual and in line with his native talent and aspiration. Then it becomes education because the worker placed in the line of work he desires, and trained in accordance with his talent and aspiration, is a growing individual—mentally, morally, and spiritually, as well as technically."[3]

7. Listen.

   This principle is included in every article and book giving advice on how to deal with people. The TWI Service wanted to make sure that "the whole story" was known before any action was taken. The JRT method includes points on how to listen to get opinions and feelings. Much as people do not intentionally do anything wrong, they also do not intentionally believe anything that is wrong. When a person says s/he believes something, s/he believes it to be true, even though it may not be. When we listen to people, we are attempting to find out what they believe to be true. The actual truth may be found later after we learn additional facts. Since the 1940s many people have studied listening and have developed ways to improve the way we listen, such as active listening and empathetic listening.

8. Don't argue.

   Here we are contrasting an argument with a dialogue. In an argument, there are two sides and each is attempting to convince the other of its own merits. This is a foolish thing to do because the "winner" of the argument will be the person with the stronger personality, and not necessarily the one who is right. Often times there is no right or wrong but rather both sides have points that both should and should not be followed. A main characteristic of an argument is that each side often does not wait for the other to finish talking before beginning to speak. This implies that the person was not listening and is in a hurry to voice his own statements. Interrupting is also a sign that the other person's ideas are not worth listening to. If one is truly trying to get facts about the situation, this is a good way to prevent that from happening. The illogic of an argument is further increased when one or both parties raise their voices. This merely demonstrates that the activity is an emotional one where the objective is to overcome the opponent instead of finding the truth in a matter.

A dialogue also consists of two sides, but here each side is attempting to understand the ideas of the opposing side. This, of course, requires a great deal of listening and feedback. The conclusion, however, has a much better chance of resulting in an agreement and even, perhaps, a compromise if required. If the statements are based on opinions and a dialogue occurs, then the worst that can happen is both parties agree to disagree.

The following seven principles align with the JRT method and will be discussed under "Using JRT."

9. Define your objective.
10. Get the facts.
11. Consider all possible actions.
12. Don't jump to conclusions.
13. Consider the timing of your actions.
14. Take appropriate responsibility.
15. Check results.

Using JRT

A supervisor's main objective is to make sure production flows through his department on time and in the designed manner. That production could be the company's final product, purchase orders, drawings or bills of material, payroll checks, or anything else the department is responsible for. Generally, the supervisor is not supposed to do any of the work that this requires, but rather to remove any obstacles his reports have that prevent them from doing their jobs. The concept of a supervisor working through people is one of the JRT foundations. The supervisor should be familiar enough with each job in the department that he understands what each person is supposed to do, and even how long that should take. Spotting a problem, something that retards production, is an art and a science. The key is to recognize changes from the norm. If a person is usually cheerful and outgoing and today you see him sullen and quiet, something may be wrong that you should know about. Supervisors are not to be social workers, solving all personal matters, but a supervisor should recognize changes that may affect production.

A problem occurring in many organizations today is that organization structures are so flat that a supervisor cannot know each job in his department. If that is not possible, then the supervisor should be able to know the employees well enough that he can understand and believe whatever they say. Sadly, sometimes this is not possible. Since a supervisor's main objective is to maintain production flow, one facet of that is to spot situations that prevent or retard that from happening. In order to do this the supervisor must know both the employees and the work they do. If a supervisor has too many reports, this will not happen and the organization will suffer. This then prevents a challenge to the proper use of JRT. The Lean term for this is Muri, which means overburden. When people are given more work than they can handle, we overburden them and the result is waste in production and stress in the individual. Although they may know the JRT method, they are not able to use it.

As stated earlier, a mistake often made in delivering the JR training is to focus on problem-solving instead of problem prevention. Both are important, but the Service emphasized, "a campfire is easier to extinguish than a forest fire." JRT teaches four categories in which problems arise. We want to identify how a problem arose so that, with practice and experience, we can push all problems into the first two categories, making the situations easier to handle.

The following are the four ways JRT teaches that problems arise:

1. Size up the situation: This is the preferred way because the situation is not yet a full-blown problem. This is an anticipation of a problem and because of that it is the most difficult. You are seeing a situation and based on your knowledge and experience, you anticipate that if nothing is done, it will be out of control and require much more effort to address in the future. It is difficult because if you need assistance, you have little to show people. They must believe you and trust in your knowledge and experience. Hopefully, others you share this with will have knowledge and experience similar to your own and they will not need much convincing. It is one thing for people not to recognize the situation as you did, but it is another not to agree that there is a problem in the making. An example of this could be an upcoming change in a policy that may sound innocuous to some, but you know that it could be troublesome to others.
2. Be tipped off about a situation: The problem does exist now, but let us say it is in the "incubation" stage. It is not significant, but it should be

addressed. Examples of this could be two people quarreling or a very punctual person not meeting due dates.
3. The situation comes to you: The problem is larger now and thus has a higher priority. It is not out of hand yet, but if nothing is done, it soon will be. An example of this could be when an employee comes to you with a request for a transfer. During a discussion with the person, you find out some information you were unaware of and find troubling.
4. You run into a situation: Now the problem is large enough that you must take care of it immediately. Perhaps you ask someone to do something and the person refuses. This is insubordination and very unlike the person as far as you know.

Identifying how a problem arose is important because we want to work to push all problems into the first two categories. As we use the JRT method, we will question how each problem arose and what we could have done to put them in a lower "cause" category.

Refer to the JRT Pocket Card in Figure 5.1 for the foundations and steps of the JRT method.

JOB RELATIONS

A supervisor gets results through people

FOUNDATIONS FOR GOOD RELATIONS

Let each worker know how he is getting along.
Determine what you can expect from each person.
Point out ways to improve.

Give credit where credit is due.
Look for extra or unusual performance.
Tell the person while "it's hot".

Tell people in advance about changes that will affect them.
Tell them WHY, if possible.
Work with them to accept the change.

Make the best use of each person's ability.
Look for ability not now being used.
Never stand in a person's way.

People must be treated as individuals

HOW TO HANDLE A JOB RELATIONS PROBLEM
DETERMINE OBJECTIVE

1. GET THE FACTS
Review the record.
Find out what rules and organizational customs apply.
Talk with individuals concerned.
Get opinions and feelings.
Listen carefully to get the whole story.

2. WEIGH AND DECIDE
Fit the facts together.
Consider their bearing on each other.
What possible actions are there?
Check practices and policies.
Consider objective and effect on the individual, the group and production.
Don't jump to conclusions.

3. TAKE ACTION
Are you going to handle this yourself?
Do you need help in handling?
Should you refer this to your supervisor?
Watch the timing of your actions.
Don't "pass the buck."

4. CHECK RESULTS
How soon will you follow up?
How often will you need to check?
Watch for changes in output, attitudes and relationships.
Did your actions help production?

Did you accomplish your objective?

Figure 5.1   JRT card.

Define Your Objective

Once a problem has been identified, the starting point is to define your objective. This is very important and sounds obvious to many people, and fairly straightforward. It is important because if you do not know what your objective is, you will not know if you have reached it. Identifying an objective sounds simple, but the truth is that most personnel situations are actually complex and have many facets. We should quantify an objective before we start, and we should also keep it in mind throughout the analysis. During this process, we should constantly question if the objective we have chosen is actually what we want to do. If it is not, we should change it and re-evaluate.

Once we have identified the objective, the four-step method is started.

Step 1: Get the Facts

This is probably the most neglected step in problem-solving. Many people get what they think is a "picture" of what happened, but either they have included their own biases or they have not gathered enough information. What we are trying to do when we "get the facts" is to build as many scenarios as possible about what happened. There are four main sources from which we can obtain facts.

First, we should check any documented records such as personnel files or any other documents. We can also include what we know to be true about the situation. Care must be taken here to separate fact from opinion. Second we should check to see what formalized rules or policies apply. These are written standards that everyone is meant to follow. However, in addition, we should also be aware of customs that are accepted in the organization. Customs are precedents from previous actions that are not written down, but are followed. Sometimes customs are more rigidly followed than are the written policies. The only way to change a custom is to turn it into a policy and then rigidly enforce it, and that usually is not an easy task. If a critical action in this given situation is a custom, and if that custom conflicts with a policy or rule, the custom will take precedence because that is what is commonly done. If the rule is to be on time for work and the majority of the employees always arrive 5–10 minutes late, the rule, in essence, does not exist. A third source of facts would be any individuals involved in or close to the situation. Again, the attempt is to separate fact from opinion by listening for the phrases, "I think" or "I believe." Try to find as much substantiation in what people say and as much correlation between people as is possible. Sometimes what people believe to be a fact is merely an opinion. Lastly, speak with the person or people at the heart of the situation. In this case you will want to get feelings and opinions because what someone says to be the truth is, in his mind, the truth. In reality, it may not be true, but you must record it as a fact because his actions will be based on it being a fact. If no one else believes it, we will question the person's actions. For example, if someone lives on the seacoast and believes the world is flat, he will not venture very far from shore for fear of falling off the edge. That will seem to be strange behavior for those of us who believe the world to be round. However, if we recognize his belief that the world is flat as a fact, his behavior is not strange at all, but rather merely cautious. We obtain feelings and opinions by following six guidelines:

1. Do not argue with the person.
   As mentioned above, arguing is counterproductive and in this situation the person may be emotional.
2. Encourage the person to talk about what is important to him or her.
   The person made some decisions and took some action based on their beliefs and interests. You have to find out what those are in order to find out the cause for the actions taken. Initially, what the person starts talking about may not seem relevant to the situation, but most likely it will be related, and you will see this relationship if given time.
3. Do not interrupt.
   When we interrupt someone, we are implying that what we have to say is more important than what they have to say. This can easily and quickly prevent someone from continuing, and his ideas, beliefs and other thoughts will be hidden from us.
4. Do not jump to conclusions.
   We are to be collecting facts and we are not to be creating any conclusions from those facts at this time. It is very tempting to want to come to a conclusion, but we must recognize that we do not have all of the facts yet.
5. Let the person do most of the talking.
   You must talk to let the person know you are interested and understand what he is saying. However, this is not the time for a lecture of any kind or an explanation of alternative courses of action. The objective is to get all the facts from the individual, so that person should do most of the talking.

6. Listen.

    As stated in the principles above, listening is critical to obtaining information. Anything you say should be only for clarification or confirmation.

You would stop collecting facts when you have run out of sources *and* you believe that you can describe a likely scenario of everything that happened. If you cannot describe a logical scenario, then you must determine where the gaps are and seek more facts to complete it. For example, if someone's behavior radically changes suddenly, that should be questioned and a cause should be found.

Note that "defining an objective" and "getting the facts" are important because they are part of the structure for good, reliable problem-solving. They require that we think through the situation and gather information. In addition, they act as a deterrent to someone who wants to make a hasty decision. This does not mean that these two actions have to take a lot of time. We can still perform them quickly. However, in performing them we are looking at all facets of the situation and collecting all possible information.

Step 2: Weigh and Decide

In this step we finalize the scenarios we made and look to see if there are any gaps or contradictions. We want to see how all of the facts relate to each other and what bearing one has on another. We must also separate facts from opinions and assumptions, and then check facts, verify opinions, and test assumptions.

We then list all the possible actions we might take. It should be noted that there is always more than one possible action, and we should list everything we can think of even if it sounds uncommon or unusual. Having listed all possible actions, we should ask five questions about each one.

1. Does the action comply with the organization's practices and procedures? Again, note that a common practice may outweigh a written rule or procedure. If the action does not comply, it is eliminated now.
2. What effect will the action have on the individual?
3. What effect will the action have on the work group? Although there are some situations that are private and technically should affect the individual only and not the work group, we should consider what would happen if the work group learned of the situation.

4. What effect will the action have on production?
5. What effect will the action have on our objective? Will it help, act as a detriment, or have no effect?

By the time we have listed all possible actions and scrutinized them with these five questions, it is very likely that the correct action is very visible to us. In any event, we must choose one action and implement it.

Step 3: Take Action

The first question to ask is who should be taking the action. Usually it would be the person who has done the analysis to this point, but that is not always the case. There are two points to consider. First, who is capable of taking action and second, who has the responsibility for taking action? If you have the responsibility, then you should take the action. But if you are not capable, then you will need assistance. Supervisors are not social workers or trained counselors. If the services of those professions are needed, they should be obtained. If you are using JRT for handling an individual, personal situation, this would be the time you may decide you need to seek professional help. In your best judgment you have made a decision what to do, but the action requires skills you do not possess. In some instances, you may not have the authority or responsibility to take this prescribed action. You must therefore go to the person with that authority and explain what you have done and request that the action be taken. If you have done a thorough job of Step 1, gathering information, and Step 2, analyzing it, the decision path leading to the action should be clear.

An important check to be made here is that the action you take is based on a fact (or facts) and not on an opinion. This check should have been done in Step 2, but the fact should be verified before you actually take action. If an action is based on an opinion and the opinion is wrong, the action will be wrong. We want to be able to say, "I took this action because of this fact."

The timing of any action is always critical. Acting either too early or too late can ruin any chances of success. The usual fault is that we act too late. Either we get busy, go on to another activity, or just do not like the confrontation that will be necessary in taking the action. Sometimes our action is based on something else occurring first, and thus we must wait a period of time. If we act too soon, we may destroy the effectiveness of our action. In any event, when we decide on an action to take, we should also decide when to take that action, and schedule it appropriately as we would any other activity. This will not only make sure that the action is taken, but also result in it being taken at the right time.

Step 4: Check Results

It is always important to follow up on our actions because our judgment may have been incorrect or other factors may have entered in the meantime. We should first decide when we should check results. We want to check as soon as possible once we know we have results, and that of course will vary with the situation. In most cases we should check a few times at intervals, and sometimes we might even have to check repeatedly. We can waste a lot of time if we check too often and for too long a time. We thus have to decide how often to check and for how long a period.

We are looking for changes in outputs, attitudes and relationships. Did your action improve the environment or make it worse? There is no need to check policies, but we should ask questions here that are similar to those in Step 2.

1. What effect did the action have on the individual?
2. What effect did the action have on the work group?
3. What effect did the action have on production?
4. Did we meet our objective?

Asking these questions forces us to look at the cause and effect of each of our actions. This is important because we must know not only why we failed, but also why we succeeded. When we started out, we defined an objective because we had to know where we were going and what we wanted to achieve. We collected facts, analyzed information, made a decision and took action. We then followed up to see if we accomplished our objective. If we did not, it would be common practice to retrace our steps and find out what went wrong. We would then repeat the method with different information and actions. If we did accomplish our objective, we should do the same analysis, although that is not a common practice. The reason to analyze our success is to discover if the success was truly caused by what we planned, or if the cause was serendipity or luck. We know the outcome was what we wanted and to our favor, but we must verify that the success was caused by what we did. Learning from this analysis will help us make better decisions in the future. Figure 5.2 shows a template that can be used when using the JRT method. A strong caution is that all copies of completed templates or notes on a given case should be shredded and discarded and not filed since they contain personal information.

Job relations training
Case studies worksheet

1. How did this problem arise?
Sizing up – tipped off – coming to you – running into you.

2. What is the supervisor's objective?

3. What are the facts?
Review the record.
Consider rules and customs.   *Listen carefully to get the whole story.*
Talk with individuals.
Get opinions and feelings.

Facts:

Figure 5.2   JRT worksheet.

Job relations training
Case studies worksheet

4. Weigh and decide

Look for gaps or contradictions.
Do the facts fit together?   *Don't "jump to conclusions".*
Consider how the facts affect each other.
Check practices and policies.
List possible actions.
Consider the objective and the effect the action has on the individual, the group and production.

Possible actions:

5. Take action

Are you going to handle this yourself?
Do you need help?   *Don't "pass the buck".*
Should you refer this to your supervisor?
Watch the timing of your action.

6. Check results

Look for changes in output, attitudes and relationships.
What effect did the action have on the person, the department, and production?
Did you accomplish your objective and improve production?

7. What foundations apply?

8. Could this problem have been prevented? How?

9. Did you accomplish your objective?

Figure 5.2 (Continued)   JRT worksheet.

Changes to JRT

JRT may have been the best personnel training for supervisors when it was launched in February 1943. Since that time, however, much research has resulted in many articles, books and methods concerning personnel relations in general, and industrial relations in particular. The JRT method has two main advantages after 70+ years. First, the material covers everything required for creating a strong, positive personal relationship. The material has been reduced to the essential concepts and thus is condensed and rich, making it easy to understand and remember. Because it is condensed, it can serve as a foundation for additional human resource training and it need not be limited to any one group. It is suitable for all employees since everyone deals with people on a daily basis. The second advantage is that this material has been "packaged" in a simple four-step method, which not only facilitates training but also creates a standard method that can be repeated throughout the organization at all levels.

Changes can be made to JRT, since much has been learned since 1943, but material should only be added and not taken away. For example, there are now programs teaching people how to listen with active listening and empathetic listening that could be attached to the JRT program. Similar courses in how to dialogue rather than argue could also be beneficial. Even programs in problem-solving would be beneficial here since JRT is really just a problem-solving program aimed at personnel situations.

Auditing and Documentation

Documentation beyond whatever is in personnel files is not required for JRT and should be prohibited. Auditing should be done by the person's supervisor.

Notes

1. Training within Industry Service, 1945. The "Training Within Industry" report: 1940–1945. Washington, DC: War Manpower Commission, Bureau Of Training, p. 52.
2. Nater, Swen and Gallimore, Ronald. 2010. *You Haven't Taught until They Have Learned*. Morgantown, WV: Fitness Information Technology, p. 15.
3. TWI report, p. xii.

Chapter 6
Selling (TWI)

> The best idea in the world is useless if it's not sold to the people who should use it.
>
> **Anonymous**

Introduction

The word "sell" can have a bad connotation. One definition is "to persuade or induce," and people associate that with a salesman attempting to get you to buy something you do not want. You just told the salesman you would buy the new car and now he wants to sell you undercoating and a protection plan. Do you need it? Another definition of sell is to cause an idea to be accepted. That definition is closer to what we mean when we say Training within Industry (TWI) must be sold. Even with a program as beneficial as TWI, people still need to be convinced to use it. They are already working a full week and have their own methods to instruct, make improvements, and deal with people. They may be satisfied doing what they are doing, and the change will take effort. Yes, they do have problems, but they have to be convinced that the TWI Programs will solve any of them. Furthermore, there are hundreds of competing programs on the market and who is to say which one is better than another. In addition, TWI was developed over 70 years ago. Isn't there something better available now? If you just found TWI by attending a conference or reading an article or listening to a friend, you may be impressed by it and think your company should be using it. The caution is to NOT try to sell it at this time. The simple reason is that you do not know enough about it and a poor start can lengthen its implementation time, or prevent it altogether. Therefore, the first person to whom you must sell TWI is yourself. You must convince yourself that the overall benefits of the TWI Programs will greatly outweigh the associated costs. This means you must study the programs and research them as much as possible. You must become knowledgeable enough so that you can counter the majority of objections that will surely be voiced.

Standard Actions

There is no overall standard plan that can be used in all companies to implement TWI, but there are some Advancing Steps to take:

1. The initiator researches and talks about TWI.
2. Managers accept a small pilot program.
3. A pilot is conducted.
4. Results are analyzed.
5. Repeat the pilot.
6. Expand.

No matter what your position is in the company, these Advancing Steps will be WHAT is done. As we know, the variability lies in the Key Points, so let us look at those. The chief executive officer (CEO), president, director, manager, first-line supervisor, engineer, purchasing agent, or operator must all prepare to convince a group of employees that the TWI Programs will be beneficial to the company, and to individuals within the company. The group must contain some members who have the authority to make financial and production decisions, and the group that must ultimately be convinced is the senior management team. Financial decisions concern hiring an Institute Conductor (Master Trainer) and accounting for employees' time. Production decisions concern acquiring production time for the pilot study. The reason that senior management must ultimately be involved is that the TWI Programs yield maximum results when they are used throughout the entire organization, no matter how large it is. In addition, everyone's budget falls under someone in senior management, and therefore the TWI activities must be justified to continue being part of that budget. Using the TWI programs will change the culture of an organization because it changes the way people think about their jobs and their co-workers. In order for an organization's culture to change, everyone in the organization must know and use the programs—from the CEO to the last new hire. The path to that team will depend on who initiates the implementation. If it is the CEO, the path will be rather short. If it is an individual contributor such as an operator, a buyer, or an engineer, the path will be longer. In this case the individual would convince his supervisor and then, together, they would convince the supervisor's peers and proceed from there.

Step 1: The Initiator Researches and Talks about TWI

The first Advancing Step is to talk to others in the company about the use of TWI. One Key Point here is that you must research the programs so that you are knowledgeable about them. You must be able to describe their pros and cons. This will be an iterative process since people will have a variety of questions, concerns, and objections. The research consists of finding the responses to the questions, concerns, and objections you do not initially know.

A main concept to remember is that the TWI Programs, individually and collectively, will improve the performance of any company, no matter how successful they are. Since the methods are over 70 years old, they have been copied and adulterated many times. They may have been changed and renamed, but the programs that are most successful are those that are based on the original methods. *If you are using Job Instruction Training (JIT), Job Methods Training (JMT), or Job Relations Training (JRT) and they are not working, that means you are not using them correctly.* The reason for this is that the TWI programs are very basic, and just one level above the main skills of reading, writing, and arithmetic (Figure 6.1). As a result, they are skills that everyone needs to know and use in order to be successful.

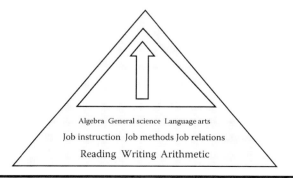

Figure 6.1 Skills hierarchy.

The work of researching TWI and finding out how it fits into your organization can be done at almost any level in the company when conducted properly. In one case, an engineer who was a member of a corporate Lean team did some research on TWI with other engineers in the community. Their common link was the local engineering association. They collectively brought me in for a demonstration at an engineering chapter meeting so they could have a better idea of the use specifically of JIT. Later I was brought back to conduct a 10-hour JIT program for several companies in the area. This was the event that convinced the companies to go forward with JIT. Note that they recognized that they would not be able to fully use JIT after this first 10-hour session. They used it to better see how JIT would fit into their companies' needs and culture. An interesting facet of the TWI Programs is that no matter how much one reads about them, one cannot really appreciate their effectiveness until one actually sees them used. At another, smaller, company, the director of HR found out about TWI and researched it. She then convinced her boss, the President, to attend the TWI Summit conference. The information he gained at the conference from breakout sessions and conversations with consultants and users convinced him to proceed with JIT.

The reason almost anyone can initiate the TWI programs is that they should not be introduced all at once, and even one program should not be introduced plant-wide or site-wide all at once. It is not enough to just say that JIT, for example, reduced a scrap rate at XYZ Company. People want to see it happen at their company. When a pilot program in a department reduces scrap by X%, people start to believe that it is something that should be considered. We are not selling with emotion; we are selling with facts. In addition, there are details to be determined in order to have the program work well in an individual company. The mechanics will be the same for all companies but there will be nuances that will vary from company to company. For example, with JIT, a method must be created to document and retrieve JBSs. One company may put them on a local server while another company may keep them in three-ring binders. People must find out what is best for them, and this is done best when the activity (documentation, for example) is used in a small group. It is easier to make changes to a process when it is smaller than when it is larger. Once all the details have been determined, the effort can spread. In addition to details of operation, the TWI programs provide a different way to think about jobs. JIT teaches a different way to think about and instruct a job, and JMT teaches a different way to think about and question a job. People will quickly see the benefits the programs offer, but it will take some people longer to get used to the different way of thinking.

An implementation will proceed faster if a small group is totally "sold" before the effort engages another group. The first group can help the second group and then the first two groups can help the others.

Change

A large part of implementing the TWI Programs has to do with change, because these programs will change the way people think about their jobs, which is a drastic change in most people's lives. It has been said that people do not like change, but I disagree with that statement. The concept of change, by itself, is neutral. People do not like whatever hurts them, and they do like whatever helps them or makes them feel better. That is, they do not like and therefore do not seek pain, but they do like and therefore seek pleasure. This is important to know, because when we want to change something that involves other people, we must find out what they really want in order to facilitate that change. If we can show someone how a change will benefit them or give them something that they want, we will not have to convince them to accept the change. They will willingly accept it. When people have been doing something one way for a long time and it is almost habitual, they feel comfortable in what they are doing. In order to get them to use another method, the benefits must not only yield better results, but they must also overcome their desire and comfort of using the old method.

Gap Analysis: Define and Quantify an Objective

Therefore, the research to be conducted requires not only finding out about the TWI Programs themselves, but also finding out what the company needs. Although the TWI Programs are training programs, they are really problem-solving programs. If their use does not focus on solving problems, they will slowly fade way. They all must be used to solve problems. That means a problem must be identified, quantified, and found applicable to one of the programs. Only then should TWI be applied to it. As mentioned in the section on JRT, if we do not define our objective, we will not know if we have attained it. This concept of defining our objective should start before we even introduce TWI to the organization. When people tell me that they want to use TWI, I always ask them why they want to. Often they do not have a clearly defined reason but only know that it can be helpful.

Finding the gap between what you have and what you need can be done formally or informally. In either case it should be quantified. For example, when I ask some people why they want to use JIT, the conversation often goes something like this:

- *Me:* Why do you want to use JIT?
- *Them:* We need to improve our training.
- *Me:* Why?
- *Them:* Our training takes too long.
- *Me:* How long does it take?
- *Them:* Job #1 for new hires takes people 4–6 weeks before they are fully productive on the line.
- *Me:* What do you think it should take or what do you want it to take?
- *Them:* We want to cut that in half to 2–3 weeks.
- *Me:* What benefit will that give you?
- *Them:* We calculate the cost of training to be $X/hour. That is the cost of salary + benefits that we are paying while we are not getting any production in return.
- *Me:* After 4–6 weeks of training, is the person fully productive on the line?
- *Them:* Not usually. It takes another 4–6 weeks for the person to be fully productive by having minimum scrap and keeping up with the other workers.
- *Me:* What would you estimate the productivity is during those first 4–6 weeks on the line?
- *Them:* About 80%.
- *Me:* So your savings could amount to {(2–3 weeks) × ($X/hour)} + {(4–6 weeks) × ($.2X/hour)} and that would be per person. Find out how many employees this would apply to on a yearly basis to see what yearly savings could be.

The same type of analysis can be applied to training existing employees on new jobs, reductions in scrap, results of cross training, and so on. Note that we do not know if we can reduce the training by half, since it has never been tried here before. This figure may have been used because some new hires have been trained in 2–4 weeks. Also, TWI literature has many examples of training time reductions even greater than that. Such data does not prove that JIT will give the same results, but it is an indication it might. This is the type of information and analysis that must be done and presented at the informational meetings. In order to find out what people's needs are, the best way is to ask them. The answers will come easily, but the numbers to quantify the needs may be harder to get. CAPA (Corrective and Preventative Action) Reports are a good source for determining needs for JIT especially. These reports document discrepancies by reporting what went wrong, why and what should be done to prevent it from reoccurring. Often the preventative action is merely listed as something to do with training, such as "operator needs training" or "improve training." Individual operators often know recurring problems that might be relieved using a TWI program. General areas to look at are training, quality, safety, productivity, and scrap or rework.

Step 2: Managers Accept a Small Pilot Program

The TWI Service recognized that senior management must be involved in order for the TWI programs to be successful. That should be a fairly obvious concept because if a department were conducting some activity that senior management was not aware of, or was not in favor of, the department could be thought of as rogue. In addition, using the TWI programs takes time and money, and senior management should be aware of both of those outlays. The TWI Service also recognized that not everyone knew how to approach management and get its approval. Indeed, this is a skill even some TWI Service representatives had not developed well. They thus created a "Management Contact Manual"[1] which was a 14-page bulletin that instructed TWI staff on how to approach and sell executives on TWI. They noted that executives wanted to know three pieces of information:

1. What the program will do for the company (quantify savings and improvements)
2. What the program is (quantify process and costs)
3. What the executive must do to make it produce continuing results (specify responsibilities)

They also emphasized that contact should be made only with the chief executive at a given site. An assistant or a department head would not be acceptable. Speaking only with the chief executive at a site is the preferable path to take, but today this is not always possible. In any event, anyone you approach will want to know the same information, even if slightly modified. The emphasis should always be on the first item: What will it do for the company and me? The CEO naturally wants to know how the company will benefit, but department heads will want to know how it will affect them.

For example, the writing of a JBS creates another document. Engineering wants to know who is going to write it, since they already write the process sheets. The documentation department (or whoever they are associated with) wants to know how they are going to be recorded and distributed. Everyone wants to know how much time this will take and where they are supposed to get the time, since they are fully engaged now.

Although you will be speaking to many people individually, there will be a time when you should schedule a meeting for a discussion that covers the three topics mentioned. This may actually require more than one meeting, since the time should be limited to an hour. The first meeting would cover the benefits to the company and each individual department. The second meeting would cover what TWI is, and the third meeting would cover the responsibilities of everyone in the company. This would include the most senior manager of the pilot program to the last new hire. If the initiator is, say, an engineer, the senior managers in the meeting may be the engineering manager, the operations manager and the human resource manager.

Meeting #1: Benefits

This is where you would cite the results of your gap analysis. It is best to identify something for each department in addition to benefits for the company as a whole. All departments can benefit from all three programs, even though the pilot may be designed for operations. All programs individually improve communication, teamwork and morale, thus reducing absenteeism and employee turnover. Citing gains from other companies is appropriate, with the caveat that every company is different, so similar gains cannot be promised here. The underlying premise is that, although every company is different, every company has some similar characteristics and problems. Refer to Figure 6.2.

Meeting #2: What Is TWI and Why Should We Use It?

Your gap analysis should tell you what your main challenges are, and therefore what J program to begin with. Some consultants encourage starting with JIT and JRT at the same time. The thinking is that the disruption caused by JIT can be overcome by JRT. I disagree with that for several reasons. First, if JIT is applied properly, disruptions will be minimal, if they exist at all. A second reason is that we want everyone to receive the 10-hour program, and so the more employees we can expose to it, the faster the implementation will go. Two sessions can be given in a day, with one in

√ Garment manufacturing in India: training time reduced from 3 months to 1.5 days while output of individual increased from 15% to 70% (compared to experienced worker). [As reported by Narayan Rao, Lean Consultant in India.]

√ Sheet steel rolling mill in Kenya: changing 2.5 ton steel rolls in mill must be done in a certain time while mill is running; novice (from accounting department) trained within two day's work as skillfully as an experienced operator. [As reported by Vinod Grover of the Kaizen Institute of India and Africa concerning Mabati Rolling Mills, Mombasa, Kenya.]

√ Manufacture of synthetic diamonds: group of novices operating in a training cell meet daily production requirements of quality and productivity after a week of training as opposed to several months without JIT. [As reported by Jed Campbell, Quality Manager, U.S. Synthetic.]

√ Manufacture of facial and body creams and lotions: efficiency increase traceable only to JIT:10%–15%; quality error reduced 60%–70% resulting in savings ~$50,000; productivity increases traceable only to JIT ~$60,000; customer and FDA audits rated as "best in class." [As reported by Adrian Oates, VP of Operations, The Autumn Harp Company, Inc.]

√ Manufacture of ceramic roofing tiles: engineering support to glazing department reduced by ~10 hours/week; glazing complaint costs reduced by 50%; rework ~0; kiln yield increased from 90% to 95%. [As reported by Jeff Lucas, Quality Manager, The Ludowici Roof Tile Company, Inc.]

√ Processing coffee beans: scrap reduced from 4% to <1.5%; machine uptime (output) increased from 60%–65% to 75%–80%. [As reported by Larry Litchfield, Manager, Keurig Production, Green Mountain Coffee Roasters.]

√ Manufacture of micro-electronics (computer chips—200 mm wafer manufacturing—IBM). The use of JIT allowed them to change from tool-centric operation (an operator runs only one type of tool) to geography-centric operation (an operator runs tools in his area). Increased operator effectiveness and efficiency. Never did it before because previous training did not allow it.

√ Manufacture of micro-electronics (computer chips—200 mm wafer manufacturing—IBM): In January 2010, had 1125 operators and technicians. Added 300+ by June (170 in May and June): "There's no way we could have brought so many people on board so quickly without Job Instruction."

√ Manufacture of micro-electronics (computer chips—200 mm wafer manufacturing—IBM) In an effort to standardize maintenance rebuild procedures, Job Breakdown Sheets (JBS) were written. While scrutinizing the procedures with the JBS, it was decided that kitting would offer advantages. Through 2010, 91 kits were created, which resulted in $2.5 million annualized savings due to better quality, fewer components used, less labor required, and lower labor cost.

√ Manufacture of micro-electronics (computer chips): JIT training of (experienced) maintenance technicians resulted in the following: reduced tool setup time, which resulted in capacity increase from 1300–1400 wafers per day to ~2000 wafers per day; cost of spare parts for given number of wafers was reduced 30%; machine downtime reduced 20%. [As reported by Bill Hill, Maintenance Manager, IBM VT, and others.]

√ Supply chain operation of micro-electronics manufacturer: processing shipping request reduced 50%. [As reported by Laura Murray, Supply Chain Analyst, IBM.]

Figure 6.2 Examples of gains from TWI programs.

√ Manufacture of ice cream: material changes on wrapping machines—failure dropped from seventeen in one month to eight in four months (17/month to 2/month). [As reported by Randy Aiken, Plant Manager, Ben & Jerry's Ice Cream, and others.]

√ Manufacturer of office furniture: moving work centers from one facility to another, were up to production within a week after the move and after one year productivity increased 50%; scrap and rework reduced 60%. [As reported by Tom Ellis, Herman Miller Production System Manager, and others, Herman Miller Company.]

Figure 6.2 (Continued)   Examples of gains from TWI programs.

the morning and the second in the afternoon. If it is decided to offer JIT in the morning and JRT in the afternoon, under no circumstances should any employee be in both sessions. The programs are not difficult but they do take some time to absorb. Being able to talk about them and being able to use them are two different skills, and the latter takes longer to master than some people think. Also, a second program should not be introduced until the first program is well underway. When I started delivering TWI, I was asked to deliver JIT one week, to be followed by JMT the following week, to the same participants. The consultant who brought me in wanted all team leads and supervisors to have both JIT and JMT, which is correct thinking. Being new to TWI and without counsel, I agreed. I later found out that the amount of material delivered was too much for the participants to retain and, since JIT was emphasized, JMT fell into disuse before it even got a chance to be used.

People often ask with which program they should start. All the J programs are beneficial and together they are even stronger. They should also be used together, but they should be learned separately. If you think of any game or activity that is composed of various skills, when you learn the game, you learn each skill individually. Principle 3 of JIT is to deliver the instruction in "chunks" because that is how people learn. When you participate in the activity, you use all the skills you know when you need them. Some companies may be very dysfunctional if they have experienced being bought and sold many times. This often results in layoffs, changes in management, and an uncertainty about even being in existence. This creates stress, and people can become very antagonistic. With fluctuations in personnel, this can result in an environment where it is difficult to concentrate on one's work, and most people withdraw in an effort to protect themselves. In such a situation, almost no program should be initiated because it will be futile. People will attend the sessions because they are told to, but they will not gainfully participate. In this case, one should start with JRT because it will give people a vehicle with which they can address the many issues surrounding them. What's more, JRT should not be limited to supervisors, as it usually is, because it will benefit all employees. I have seen only one company in that condition, but I know there must be others in a similar situation. Others would probably not be in a position to seek any training, thinking they have larger problems to focus on. This company engaged a consultant to help them back onto an "even keel," and thus he brought me in. It was very beneficial and did help to turn around the situation. Since this is a rare occurrence, the first TWI J program to start with should be JIT. The reason for that is that JIT leads to standard work. Production (in every department) must be standardized before it can be improved. If work is not standardized and people are doing a given job various ways, it is difficult to identify problems and make improvements. Once JIT has been used to the point where people are comfortable with it, and it is returning dividends, JMT should be introduced. JRT can be introduced at any time, but when it is introduced, the participants should not be engaged in any JIT sessions.

A presentation of what TWI is and why a company should use it could be a volume in itself. The Key Point is to suit the presentation to the audience. Everyone has something that they care about and the TWI J Programs have something for everyone. The objective is to match them up. A demonstration is helpful here, but if you have not been through the 10-hour session, it is not recommended. A description of the programs and the four steps of each one will give people an idea of them. The main objective is to show how the J program(s) will meet the company's needs. I will include some bullet points here and you should expand on them with other material you find. The points marked with asterisks (**) are discussed further in Appendix III.

- Always include quantitative examples of gains such as those shown in Figure 6.2
- Review the four-step methods of each of the J programs
- Include the JIT motto
- Safety training is included with production training
- The TWI programs are practical and easy for everyone to learn and use. They contain good management principles, so as a result, everyone will be following these principles
- The J programs are necessary because:
    - JIT: Performing a job and instructing it are two different skills
    - JMT: Many people do not know how to vet, sell, and implement their ideas

- JRT: Many people do not know how to handle difficult personnel situations consistently
- Each J program engages ALL employees, resulting in improved cohesion among employees
- JIT enables standard work (you cannot have standard work without JIT)
- JMT enables a real continual improvement program: Involving everyone
- JRT enables respect for the individual
- Standard work reduces variation so improvements can be identified and made
- JIT resolves many of the common instruction problems (Figure 6.3)
- All J programs improve communication, teamwork, cohesiveness, and morale and reduce turnover and absenteeism
- All J programs change the culture positively so people think more about their work
- ** J programs align with Self-Determination Theory[2]
- ** J programs are a foundation of a Learning Organization

#	Where used	Technique
1	Step 1	Check physical ability—confirm capability
2	Step 1	Check desire–confirm wants to do job
3	Step 1	Check prerequisite knowledge—confirms qualifications
4	JBS	Give only relevant information—only what is necessary to do the job
5	JBS	Be concise but understandable—facilitates understanding and retention
6	JBS	Include all details—for required Q,P,S,C
7	Steps 3 and 4	Repeat as needed—retention
8	JBS	Give information in order—facilitates understanding and retention
9	JBS	Use known terminology—avoids confusion and intimidation
10	JBS	Use the same words—facilitates understanding and retention
11	JBS and step 2	Deliver in 'chunks'—facilitates understanding and retention
12	JBS	Explain why—increases interest and understanding
13	Steps 3 and 4	Check real understanding—no assumptions = fewer errors once on job

Figure 6.3 Job instruction training: techniques for proper instruction enabled through the JIT method.

It would be beneficial to conclude the meeting with a short discussion to find out what people think about the programs. Emphasize that you are not asking them to make a commitment at this time, but just want to know whether or not they are in favor of TWI, and what their reasons are for that thinking. A pertinent question is what program would they choose to use first IF they decided to employ TWI. Say that you would like to meet again to share an implementation plan proposal with them.

Meeting #3: Responsibilities and Commitment

Now that this group knows how the TWI Programs can help them, and what they are, it is time to explain their responsibilities in the implementation process, and the commitment they must make to achieve it. You have done most of the work necessary to prepare for this meeting but some still needs to be done.

JIT is usually chosen as the first program to use because it standardizes production. If JMT is used first, you will find a need to standardize processes and instruct employees as the JMT program progresses, which is another reason JIT should be used first. If some people have a strong desire to start JRT with JIT, the stipulation must be that two different groups will be required. One group of employees will receive JIT and a different group will receive JRT. The most important consideration in choosing the first program is based on the greatest need. JIT would be chosen if the greatest need is standardization,[3] while JMT would be chosen if the greatest need is the development of a continual improvement program. JRT would be chosen first if the organization is dysfunctional to the point that either of the other two programs would be difficult to administer.

You must now decide who will receive the 10-hour program, based on the situation at hand. If JRT has been chosen, you will know who the people are who will benefit most from the training. In addition, their supervisors and, if possible, their supervisors, should be included in the training. In JRT, one's supervisor should act as a coach to guide the individual with the process as needed. HR management should be included and they can also act as coaches.

If JMT has been chosen, having some idea of what you want to accomplish will reveal who is to be trained first. Choose a part of the production area and deliver the 10-hour session to the line and staff in that area.

JIT is most commonly chosen first and you must select a team to work on the pilot program. The people selected would be those who work in the chosen area. One client had quality problems in the packaging/shipping department, so she decided to start there. Another had problems in the glazing department, so he focused on that area. Once you have chosen an area, the 10-hour participants would include operators, team leads, supervisors, support engineers, quality personnel, and maintenance personnel. Employees must be chosen from all shifts. There will be a maximum of ten people per group, and there can be two groups for a total of 20 participants. Most likely you will want an employee to be developed to deliver the 10-hour JIT program, so some thought should be given to a trainer candidate at this time, but it is not necessary to make a decision now. Having selected a list of possible participants, the jobs that are to be improved should also be identified at this time. Also, depending on the size of your organization, you might want a coordinator who is separate from the trainer. Someone must drive the program. In many companies, one person is selected as a JIT trainer/coordinator. As the program grows throughout the organization, trainers are added, and the original coordinator restricts himself to that activity. Also, one or two of the participants should also be identified as coaches. These would be people who will assist the trainer in coaching participants after the 10-hour session ends. There is no need to identify these people now, but you should be aware that you will need them later. The final team will consist of the 20 participants. From those participants, there will be—one or two coaches, one trainer, and one coordinator (and those last two positions can be combined). Finally, the team should have a sponsor. This would be a senior manager whose job it would be to remove obstacles for the coordinator.

You will need to hire a TWI Institute Conductor (Master Trainer) to deliver the first two sessions at least. I have a client who is on the continual improvement team, and her boss gave her the assignment of implementing TWI. She said she spent about a year trying to do it herself without much success. She focused on JIT, had the original Trainers' Manual and some other references, but she was not doing it correctly. She knew she needed JBSs but was still unsure how to write them. Although this book would have helped her, I would still recommend using an experienced JIT trainer for the first session or two. It is very much like telling someone to learn how to swim, and then giving them a book on how to swim. A main principle of TWI is learning by doing, and that must be done with correct feedback. There are many TWI consultants available. Most are or were Lean consultants, and they picked up TWI training to broaden their offering. It would be good to have a central, independent agency train people in using the TWI Programs and then certifying them. That would make the job of selecting a TWI consultant much easier. Because we live in a competitive environment, and a real certification process is expensive, TWI certification programs do not exist in any country where the program is not under government control. That is not to say that consultants do not "certify" clients as TWI trainers, but one must look at the origin and thus the value of the certification. As a result, when choosing a TWI consultant, contact their past clients and find out as much as you can from them before making a decision. Obtain a quotation so that you can determine expenses.

The agenda for this meeting would be to first show the rough plan you have developed. That would include

- A description of the area for the pilot, why it was chosen, and expected gains
- A list of possible participants
- The duties and responsibilities of the trainer and coordinator
- The duties and responsibilities of the sponsor
- Time required for all employees (10 hours of training and instruction time)
- Cost of employees' time
- Cost of consultant

Explain that this is a pilot study because we must actually use JIT in order to really understand what it will be able to do for us. If we believe it to be beneficial, we will continue implementing it throughout the entire facility. Also, this is not a "one-time" training plan but rather a program that is intended to alter the company's culture. We intend this to be embedded into the operation of the company to the extent that it will be the way we operate and the way we think about operating. If we decide it is not beneficial,[4] we will stop the program completely. The plan does not have to be entirely complete at this time. You just want to be able to demonstrate what will be gained for the cost (time and money) outlay, and how you will do it. An experienced consultant can give you any estimates you may be missing. Be conservative with the time estimates. Before leaving the meeting, get a commitment from everyone that they will support this effort and complete any task for which they have a responsibility. Everyone from the CEO to the first-line supervisor in the pilot study should understand and accept the tasks and responsibilities for their position. Some companies reduce this to a written contract that everyone signs. At a minimum, you should have a commitment on

- The plant area on which the pilot program will focus
- Quantified parameters that you wish to improve
- Expected quantified results
- The participants in the first section of training
- The duties and responsibilities of line management who will be involved.
- The sponsor: Duties and responsibilities
- A timeline
- Costs

Once you do have commitment, you are now ready to prepare for the implementation.

Notes

1. http://www.lean.enst.fr/wiki/bin/viewfile/Lean/TrainingWithinIndustryEnFrancais?rev=1;filename=Management_Contact_Manual_1944.pdf, p. 2. Note that this manual contains some helpful information, but it cannot be used as written because structures have changed since 1944. The company, not the government, now pays for the training and there is no certification process so each consultant must be vetted.
2. http://www.selfdeterminationtheory.org.
3. Note that many companies have work instructions or standard operating procedures (SOPs) and the people in charge of them believe that the company already has standard work. It must be brought to their attention that they have standard procedures, but not standard work, because the operators do not follow them. The selling point here is that JIT will be used for getting the operators to follow standard procedures.
4. Note that if it is decided that JIT is not beneficial, it was not used properly. However, some people do not wish to pursue TWI because they do not want to incur such a change at this time.

Chapter 7

Implementation

You can't build a reputation on what you are going to do.

Henry Ford

Preparation

Finalize Plan

You have received approval for the TWI pilot, so now you must finalize the plan, prepare everything, and then begin executing the plan. You should first engage the TWI consultant. You have done research on those available and now is the time to make a choice. This will result in a firm date for the training, and the consultant can also help you with some of the preparation details.

General

Now is the time to select the people who will be the coordinator, the trainer and the coaches. Appropriate job descriptions can be seen in Figure 7.1. Note that when starting out, the coordinator and the trainer can be the same person. In fact, at this point it may not be a full-time job. Since the person who fills the coordinator and/or trainer position is currently gainfully employed, this is not too late to start withdrawing that person from his existing job by off-loading some of his duties in order to enable him to assume some new duties. This, of course, is easier to see if the person

TWI coordinator
Possible tasks

1. Schedule and prepare for the 10-hour sessions
2. Schedule participants
3. Maintain history of participation
4. Maintain training matrix
5. Follow up with supervisors after the training to see how it is being used and to offer input as needed (includes helping set up an audit program)
6. Obtain and distribute results
7. Assess JBSs, JIT instruction, JMSs
8. For JIT and JMT, oversee creation and documentation of JBS and JMSs
9. Coach creation of JBSs and JMSs
10. Coach instruction
11. Deliver the 10-hour programs if they have been trained to do so,
12. Facilitate JBS groups (for getting consensus), JMS groups for getting consensus on larger JM improvements, and JR groups
13. Help people identify problems that can be solved or reduced through the use of the programs, etc.

Qualifications

1. Several years with the company
2. A broad knowledge of all departments in the company
3. Interest in the J programs and desire to be coordinator
4. Participated in at least one of each of the J sessions

TWI trainer
Tasks

1. Deliver the 10-hour "J" program
2. Maintain participant's manuals

Qualifications

1. Desire to be a trainer
2. At least a year with the company
3. Participated in at least one 10-hour program
4. Observed at least two 10-hour programs
5. For JIT and JMT, is a coach
6. Delivered at least one 10-hour program

JIT or JMT coach
Tasks

1. Assist people in using JIT or JMT

Qualifications

1. Participated in at least one 10-hour session
2. Participated in at least 15 JBSs or JMSs
3. Desire to help others use the JIT or JMT method

Figure 7.1 TWI job summaries.

is also the trainer. With Job Methods Training (JMT), and especially Job Instruction Training (JIT), just before the first training week it is highly recommended that this person be solely dedicated to the new position as coordinator/trainer. Specific people do not have to be selected as JIT or JMT coaches at this time, but everyone should be aware of the need for them, so people's interest can be assessed. In one medium-sized company of a few hundred employees, the vice president (VP) of operations was my first contact, and he brought me in to deliver the first JIT session. He stayed in the coordinator position until the program was up and running, and we found that worked very well for several reasons. First, because of his senior management position, he could make sure that the program got the attention it deserved. All employees would recognize its importance. Second, as employees started using JIT he was in a better position to select employees to be in the trainer and coach positions.[1] He could also make sure that the JIT program was structured and used as he wanted.

You know the area of the company that will be used for the pilot program, so you should briefly meet with each member of line and staff management for this area, and answer any questions they may have.[2] Team leads should be in the first training group, and it is helpful that they know the background of the pilot study. At least one first-line supervisor should be in the training, but all first-line supervisors should know about the TWI programs, so they should be contacted and briefed. They will be "covering for" the supervisor who is in the training and they should know why. It is also helpful to let them know why they are not in the training. They could be invited to be observers, but that is usually not possible, since one supervisor will be away for two hours already and production needs to continue. Depending on the size of the organization and its structure, managers, directors, and other senior management may or may not be in the training. Whether they are or are not, they should be informed and briefed on the event. The sponsor of this program should also be briefed, but in addition s/he should be kept up to date on your progress on a regular basis.

Reviewing what is going to happen with all participants involved (those who will attend the 10-hour session) can be done in several ways, depending on your individual situation. One method is to talk with each person individually. This will take less of the person's time, but much more of your time. One disadvantage of this is that someone may ask a question that you did not think of. If you feel it is important enough for everyone to know, you would have to retrace your steps and bring the others up to date. Another method is to have one or more orientation meetings where you can briefly tell everyone about what has happened to date, what the training is, and how it will be conducted. You can describe some jobs that will be used as demonstrations in the training and also ask for their suggestions. You can tell them why they were selected and why others were not. Emphasize that, once the TWI program has been approved, everyone will receive the 10-hour program. Tell them what is expected of them during the training week and afterwards. If possible, have a member of senior management introduce this orientation meeting, so that everyone realizes that the company takes it seriously. Some companies have created a TWI orientation slide deck, to be shown during this meeting and also to new hires. During this meeting, you should tell the participants that everyone is expected to attend each day's session, so no days off are to be scheduled for that time. (You have already selected a week when there are no scheduled holidays.) It is important for 100% attendance because most people will learn something every day of the week and the more they see of the TWI program, the better they will understand and use it. At the end of the meeting, distribute a letter similar to the one in Figure 7.2. This will cover most of what you have told them and can serve as a reminder. Note that the best time for this meeting is about 2–3 weeks before the scheduled training. If it is conducted much before that, people may forget about it. If it is conducted too much after that, people won't have time to prepare themselves mentally and physically. You also want to give the correct impression that this is a planned effort and not a hastily set up "time filler." Let them know that they can contact you in the meantime if they have any questions. An orientation such as this is another opportunity to determine if there are other objections or obstacles. Such obstacles include the inclusion of employees who speak the company's language (e.g. English) as a second language or who are deaf or illiterate. They must be included in the training, but the decisions of when and how to include them must be addressed for success. You should identify and respond to any challenges that come up, and you should also inform people of those that have already been discussed previously.

Job Selection and Training Room

At this point you can take a closer look at the jobs to be used for the demonstrations. There will be 20 participants in the first week of training, and each participant must demonstrate the TWI method. That means you will need 20 jobs that can be used during the training. The Job Relations Training (JRT) participants need only think about what their case study will be; no further preparation is required. It is usually best to perform the JIT and JMT demonstrations in the training room. Each JIT demonstration should take about 15 minutes to perform, and each JMT demonstration should take 20–30 minutes, so you must select jobs based on those criteria. The consultant can help you select jobs for this first week. Many people

Job instruction training at
Your company
Somewhere, USA
April 30 – May 4, 2016

Introduction to participants

Don Dinero, institute conductor
April 22, 2016

To the JIT Participants:

It is a great pleasure for me to introduce you to our upcoming 5-day training program on **Job instruction.** This program is one part of the Training Within Industry (TWI) series that, for over 70 years, has been helping employees such as you improve their performance by developing a well-trained workforce.

From **Monday, April 30 through Friday, May 4, from xx am to yy am or xx pm to yy pm,** you will be learning how to train people to quickly remember to do jobs correctly, safely, and conscientiously. Using this method, you will have

- Less scrap, rejects and rework
- Fewer accidents
- Less tool and equipment damage
- Less employee turnover
- A basis for standard work, and
- A basis for continual improvement

This program is for anyone who is charged with instructing someone in a job but it is also useful for everyone who will be transferring knowledge to others, i.e. everyone.

Preparation:

One of the key tenets to this training is an on-the-job component that immediately gets participants using the method on the real-life jobs they do and/or oversee. The purpose of our training is for you to learn the four-step method of Job Instruction and the very best way of doing that is for you to actually apply it—to learn by doing.

Each of you will have a chance to bring in a job for practice during the program. After Tuesday's training, I will be helping three of you prepare for your demonstrations on Wednesday, and so forth. In the meantime, please review these notes and be thinking about a good job to use for your demonstration.

1. Review your job areas and pick a job that:
   - Is an actual job done at your work site
   - Can be done in about a minute or less
   - Is simple and easy—do not choose something that is a trick or a puzzle

   Some examples others have used in these sessions would be:
   - Light assembly work – e.g. installing a terminal on a wire
   - Taking a measurement of a completed part
   - Adjustment of a fitting or part
   - Inspection of a part or completed product
   - Packing of parts into a container
   - Filling out a form (hardcopy or on computer)
   - Packaging of instruction manuals or other paper materials
   - Proper usage and fitting of safety tools like ear plugs or a mask
   - Writing a report
   - Writing a requisition or purchase order
   - Scheduling – production or a meeting

2. Because we plan on conducting the sessions entirely in the training room, select a job that can be demonstrated off-line and that you can demonstrate in front of the group. If your work consists of using large machinery and parts, try to find some real job that is portable (for example, how to use a measuring instrument).

3. Since the method calls for you to actually do the job several times, plan to bring enough parts and materials to do the complete job at least 10 times.

Figure 7.2   TWI letter to participants.

4. Bring to the training session any other materials, tools, or equipment that are actually used in doing the job. This includes safety equipment such as glasses. **You will not need these materials on Monday.**
5. As mentioned, I will work with each of you on a 1:1 basis to help you prepare for your demonstration. This coaching time usually takes about an hour, so **please budget for an extra hour** during the week in addition to the approximately 10 hours of training.

During the training, you will be asked to demonstrate for the group how you would train somebody to do this job, using the job instruction method you will be learning. Each of you will also be asked to play the role of learner for another participant demonstrating his or her job so that, when you teach your job, there will be a person who is actually learning it.

You will learn not only by demonstrating a job, but also by observing others demonstrate. Because of that and because the schedule is fairly tight, we ask that you:
- **Be prompt**
- **Attend every session**
- **Notify appropriate people of this training so that you will not be interrupted during the session**

Job Instruction Training is primarily a problem-solving tool and it is very powerful because many of the causes of problems can be traced to inadequate training. Therefore, think about the obstacles that prevent you from doing your job well, since most of these problems can be solved through adequate training. That is, they can be used to help determine where Job Instruction Training should be focused.

Contact:
If you have any questions at all about the training, please do not hesitate to contact me. If you do not get through tome directly, leave a message with your question and a phone number and I will get back to you as soon as I can.
Work phone: 585-453-9478
Cell phone: 585-305-6820
Work e-mail: dadinero@TWILearningPartnership.com

Closing:

I am very much looking forward to working with you on this TWI program. This program has been used around the world for many years and it has had tremendous effects. It has been referred to as "the most underrated achievement of 20th century American industry." It will be my great pleasure to pass it on to you.
See you on April 30!

Regards,

Don Dinero

Figure 7.2 (Continued)   TWI letter to participants.

who are unfamiliar with the JIT method have a tendency to choose jobs that are much larger than are usable for a single JBS. The focus of the jobs should be on the desired improvements and expected results, if at all possible. Selecting demonstration jobs before the training starts saves a lot of time during the training week. You know the area cited for improvement, and, with the help of the consultant, you can choose jobs that are of the appropriate size and type that will be most conducive to demonstrating in the training area. If this information is shared with the participants, it will also give them some time to think about their demonstration in the context of JIT. See Figure 7.3 for an example of a demonstration assignment matrix. It

Demonstration jobs and coaching schedule

Group A	Monday	Tuesday	Wednesday	Thursday	Friday
7 am – 9 am		Training for group A			
9 am – 10 am		Chris S. – 5S tagging	Billy K. – Gowning	Gil J. – Solder sensor	
10 am – 11 am		Clark L. – Calc. Takt time	John N. – Write NC report	Matt K. – Log in to operation	
11 am – 12 pm		Will M. – Clean and pack assemble	Roz C. – Stopper tensile test	Kraig T. – Install monitor arm	
12 pm – 1 pm		Lunch			

Group B	Monday	Tuesday	Wednesday	Thursday	Friday
1 pm – 3 pm		Training for group B			
3 pm – 4 pm		Wally B. – Inspect locks	Vinnie G – Assemble roller	Stan A. – Write ECO	
4 pm – 5 pm		Carl K. – Line clearance	Mike I. – Enter PO	Lucy K. – Rivet lower stage	
5 pm – 6 pm		Bill J. – Assem. transducer	Larry S. – Set up UV machine	Sam Z – Inspect trailer	

Figure 7.3   Typical schedule.

is important that the JMT demonstrations are for improvements that have not yet been made or even started. A person can have an idea of what he wants to improve, but the idea in JMT is to analyze a process and devise improvements using JMT. Note also that the demonstration tasks should be related to the participants' jobs. We want to make it very clear that JIT or JMT not only works, but also is applicable to whatever each participant does at the company. Efforts can be made later to show how these programs relate to activities outside of work. Having a work-related demonstration also facilitates implementation of the program.

The training room should be large enough to seat ten people, as shown in Figure 7.4. This set-up enables discussion and involvement. If space is available, observers can be present. Criteria for the training area are

- It prevents or reduces distractions (consider noise, room temperature, and lighting).
- There is enough room for everyone to be seated comfortably.
- There is enough room to perform demonstrations.
- A whiteboard (4 inches ×6 inches) and a flip chart are available.

The result is that neither a conference room nor a standard training room is required, although that is what is usually used. If such a room is used, that usually will preclude the use of compressed air, chemicals and most equipment and machinery for the demonstrations. It is best if the demonstrations are performed in the shop, where the jobs occur, because everyone

174 ■ *The TWI Facilitator's Guide: How to Use the TWI Programs Successfully*

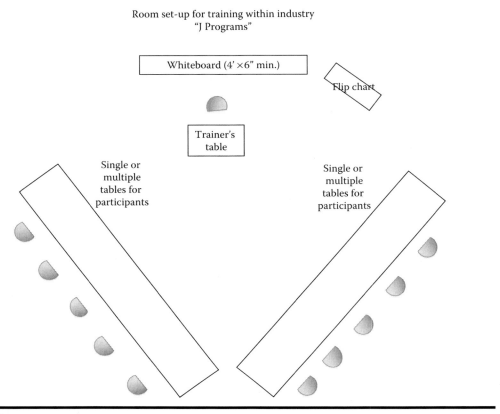

Figure 7.4 Typical room set-up.

can get a better appreciation of them in their regular setting. However, this is not usually possible but often the demonstrations can be modified with simulations to work around this obstacle. There are instances where it is possible to use the actual workplace for demonstrations. One participant wanted to demonstrate instructing a person to use a modified electric floor jack to stack 55-gallon drums. He scheduled his demonstration when there would be no distracting activity in the receiving department. As a result, the demonstration was conducted and there was enough space for everyone to see, and it was quiet enough for everyone to hear. This is important because participants will learn both from their own demonstrations and from watching others. Another client is a meat processor and most of the jobs had to be conducted in a refrigerated area. A refrigerated area that was set apart from the main processing area was chosen and many of the demonstrations were conducted there.

Make sure that the room or area is reserved and signs are posted, if necessary. Also, make sure the room environment is conducive to training.

That is, it's not too hot or too cold or too noisy or dirty. It is also less disruptive if the same room or area can be used throughout the week. It is usually a very busy week and moving the training from one room to another is a time-consuming activity that should be avoided if possible.

Documentation, Assessments, and Publications

Having done all of the above, you are now ready for the TWI training. However, there are two subjects that you must start to think about now because they must be accomplished shortly after the end of the week's training. Concerning the JBS creation process discussed in Chapter 3, you must now think about the details of the process. Some of these details are

- Who will write the JBS first draft?
- How will you get consensus?
- How will JBSs and JMSs be documented and controlled?
- How will JBSs be updated?
- Where will the instruction be done?
- Will signoffs be required for JBSs?
- How will JBSs affect existing work instructions?
- Who will be the participants in the next 10-hour session?

Concerning JMT, there is less preparation, but you must consider:

- Who will maintain the documentation and publication of the MBSs?
- Who will have access to them?
- Is a separate implementation plan needed to make changes? If so, what is it?
- Will signoffs be required for MBSs?
- How will use of JMT affect an existing CI program?
- Who will be the participants in the next JMT session?

The only preparation for JRT is to determine how to assess the use of the method after the training. Each supervisor will coach his or her reports, and records will not be kept unless it is required for Human Resources (HR).

In addition to these factors, an important function that is not mentioned in the original manuals is that of assessment or audit. Conducting an audit is necessary whenever we wish to change behavior. The two main reasons for this are that we may have trained the individual improperly, and/or the individual may have lapsed into old habits before new ones could be formed. With JIT, we should check to see that operators are following the JBSs they have been trained to. We should also check to see that the instructors are delivering the JBSs with "JIT speak" as noted before. A coach or the coordinator should also periodically sit in on a 10-hour session to verify that the trainer is following the program. With JMT, the MBSs should be reviewed periodically with the person or people who wrote them to verify that the method was followed. With JRT, there is no breakdown sheet and all information is confidential. There is also no list of jobs that should be instructed or could be improved. However, a periodic meeting could be scheduled where the JRT participants review and critique some of the situations that have been addressed with the JRT method. All assessments, including this one, should be based on an educational purpose and not a punitive one. People should be educated so they use the method correctly and not be chastised because they used it incorrectly.

Training

Points Common to All Three J Programs

We have already discussed why the five-day schedule optimizes learning. Just as it takes time for a dry sponge to absorb water, it takes time for the knowledge in a J program to be absorbed by people. Having some time away from the training is critical for people to reflect on what they have seen and heard. Remember to use JIT when delivering the TWI programs. Two of the four points to get ready to instruct are first to prepare the right material, equipment and supplies and second to properly arrange the workplace. Thus, make sure the training area and all materials (such as participant's manuals[3], name tents, etc.) are ready. The use of a whiteboard is highly recommended. The only alternative would be a flip chart if a whiteboard is not available. The disadvantage of a flip chart is the size and that it is not erasable. An advantage is that the papers can be posted around the room as reminders during the week. A good use is a combination of a whiteboard with flip charts. Certain phrases such as the definition of the J program being delivered, or the JIT motto, can be on a flip chart and thus referred to throughout the week. In no case should computer slide decks be used. Computer programs such as MS Power Point or MAC Keynote are better suited to larger audiences or where sophisticated charts must be displayed. They have no place in training a group of 10 people because the participants will be drawn more to the slides than the trainer. It is also easy for the trainer to rely on the slides as a "crutch" and depend on them conveying too much of the information. When a trainer writes on a whiteboard or flip chart, s/he can create a stronger connection with the participants and the training will be more effective.

Let us take a look at each of the five days of training. This will not be a complete description of the training in the five days, but rather mainly the areas that I have added to material found in the original manuals. The original manuals can be found on the Internet and should be followed here. We will start collectively because there are some comments that apply to all three J programs, but then we will discuss each J program individually. The following patterns follow the original training manuals, but I have modified them to satisfy some needs that have arisen since 1945. A company trainer should also modify them to better fit his or her own company, but in no way should they eliminate anything that was in the original manuals. In addition, the named principles should also be included. The Key Point for making changes is to follow the Scientific Method. Perceive a need; propose a solution; make a change; check results. If the change meets the need with no deleterious affects, continue with it. If the change does not meet the need, or there are some poor effects, return to the original state and propose another solution.

The original manuals were written with the intent that most, if not all, of the participants would be first-line supervisors. The reason for this is that the country was, as they often said, in an "emergency situation." Because of the war, everything was in short supply and the TWI Service did not have enough personnel to train everyone in an organization. They believed that first-line supervisors were the correct focal point to apply the training in order to get the largest payback. They did recognize, however, that these programs would be beneficial for everyone and thus let it be known that if space were available in a session, employees who were not supervisors should attend. They also believed that supervisors should be the instructors, which is often not the case today.

The JIT Trainer's Manual has the phrase, "Work from this outline—Don't trust to memory" at the bottom of each page. JIT was the program used first, and the one used the most, and when they were finalizing it, they discovered a problem. Many of the first people to use JIT were professional trainers and teachers, and thus they personalized the training. Because they were professional trainers, the outcomes were good, but when the TWI

Service audited several groups, they found out that they were all different. A standard program was required that could be used throughout the entire country, so the above mentioned phrase was included in the manual. Today some people have interpreted that phrase to mean that one simply reads the manual, but that is not true for any of the J programs. The only exception is one case study in JRT that the trainer is told to read aloud so that no details are missed. Here, the warning is to make the reading animated in order to keep the attention of the participants. The JMT Manual conveys the intention of not reading from the manual.

> Paragraphs in quotation marks are to be presented either by using the exact words of the text or expressing the exact meaning in the Trainer's own words. In case of the latter, special care should be taken to convey the exact meaning every time.
>
> Wherever the expression "(some discussion)" appears, there should be brief discussion to make the point clear or to reach agreement with the group. Words in bold face are key words, which provide the Trainer with a quick clue to the statement made in the sentence.[4]

A letter from C.R. Dooley, the director of the TWI Service, addressed to "The War Production Trainer" was included in each manual. The letter in the JIT manual stated:

> To assure a uniformly high standard on a Nation-wide basis, you should work from this outline ALWAYS. Don't deviate from it. Don't trust to your memory, regardless of the number of times you may present the plan. It is not difficult and if you follow the instructions you can't fail—furthermore, you will find it an interesting job.[5]

If at all possible, a member of senior management should introduce the first day of all TWI Programs. This will serve to demonstrate the importance of the training and that it is not a "flavor of the month" exercise. Once the general introduction is done, the trainer should continue with introductions of himself and all participants. This is when administrative matters are discussed such as silencing cell phones, and so on. As a "warm up" exercise, I always supply a quote for each day. It gets people's minds working and also emphasizes the general topic for the day. For example, for the first day of JIT, I write and then discuss a quote by Charles R. Allen, who was instrumental in the development of JIT, and who oversaw the training of shipbuilders during World War I.[6]

> The trouble, of course, lies in the fact that whether whoever gave the instruction was or was not a first class man on his job, he did not know how to "put it over." He may have known his own game, but he did not know the instruction game.[7]

This quote reveals that doing a job and teaching someone how to do that job are two different skills, and the latter is what we will be learning this week. On Tuesday, the quote, which is attributed to Sophocles in 445 BC, is "One must learn by doing the thing; for though you think you know it, you have no certainty until you try it." Learn by doing is a fundamental principle of the TWI Programs and that is why everyone must do a demonstration.

Each of the manuals started the same way. The trainer was to put the participants at ease and to avoid giving the impression of a classroom. He should advance the idea that this is a meeting where discussions will take place and people will learn. The learning will not be about people's specific jobs but rather aspects of those jobs that will assist them on a daily basis. This would then lead into the "Five Needs of a Supervisor." With the introduction of Lean and other such programs, many people are familiar with the four metrics that management uses to monitor the company's performance. Some have added additional categories, but the four original ones were quality, cost, delivery and safety. I start each session discussing these so that I can tie the specific J program to the company's goals. I have also changed "safety" to "personnel," which includes safety, teamwork, and morale. Getting people to identify the four metrics is usually difficult even though posters, bulletins boards, and metric charts can be seen everywhere. That leads me to question what employees really know about it. To get the participants to tell me the metrics, I switch the discussion and make them the consumer instead of the producer. I ask what they think about when they want to make a large capital purchase, such as a refrigerator. They finally tell me they want quality at a reasonable cost and they want it at a specific time. That matches what their customers also want. The discussion of personnel is brought up because someone has to do the production. Finally, we call this a balanced scorecard because we must pay attention to each metric or we will run into problems.

The discussion of the four metrics would lead to the question of what supports the metrics. In JIT, standard work is known to be a main requirement. What is standard work and why is it important? I define standard work simply as a given job being done the same way, every time it is done, no matter who does it. I use the acronyms: SJ, SW, ET, EB, meaning same job, same way, every time, by everybody. We are not concerning ourselves with the time it takes to do the job and we are assuming we will have the proper inventory and materials to do the job. This effort is concerned only with instruction. As with any skill, one must learn the skill first and only then should the person work on increasing his speed.

A difficulty often arises because many companies have standard processes and thus they think they have standard work. The difference between standard processes and standard work is that processes are what is written while work is what is done. Many people, especially those who write the standard processes, get frustrated and upset because they do not understand why operators do not follow their processes. The simple reason operators do not follow standard processes is that they are not able or it is too difficult to understand what the process is from the document. Sometimes the job physically cannot be done as described. It is difficult to explain even a simple operation in text, and thus the instructions are so complex they are incomprehensible. Also, standard written processes have a tendency to grow over time. When a problem of some kind occurs, a correction is added to an existing process to prevent the problem or mistake from reoccurring. However, items are usually never removed from a process document in fear that their removal may cause a future problem. The second question of why we should use standard work is usually more difficult for the participants, no matter who they are.

Most people believe they should be doing standard work, but getting them to tell you why is not always easy. We want to use standard work so that we can more easily see the variability in what we do. It is easier to catch a mistake if all you have to look for is something that is different. If many things are different and only one thing is wrong, the selection is much more difficult. So now we realize that we can use standard work to identify our mistakes faster, but the question that remains is how do we get standard work? That is, how do we get people to follow the standard processes?

This is the lead-in to the Five Needs of a Supervisor. The original JIT Trainer's Manual defines a supervisor as "a person who directs the work of others."[8] This was included because it was recognized that not everyone in the group might have direct reports and such a person might think that this training did not apply to him. I mention Five Needs of a Supervisor, but I also emphasize that all employees have these five needs and if everyone mastered all needs, they would be successful in whatever they do. The discussion of the five needs ends with the need of knowing how to instruct. Knowing the JIT method satisfies that need. The summary to this point is

- We all want to improve the four metrics of quality, cost, delivery, personnel.
- One concept that will help us with that is standard work.
- The only way to get standard work is with standard training and the only standard training is JIT.

The JMT session starts out the very same way. The stage is set with a discussion of the four metrics and the same lead-in question is asked. How do we support these four metrics? If the participants have received JIT, they may say standard work, but there are other activities we must use for our balanced scorecard. Continual improvement (CI) is one of them. We first define continual improvement as ever-increasing productivity, and we define productivity as the rate at which an organization produces goods and/or services in relation to the amount of material and effort required. Then the question becomes, how do we get continual improvement? We go to the five needs discussion and this time JMT is the last need we talk about. First, it must be agreed that everyone has ideas. They may not all be good ideas, but everyone does have ideas. What many people do not know how to do is vet their own idea and determine by themselves whether it adds value to what we are doing. Once we know how to vet our ideas, we must know how to sell them because if we cannot convince people to use the idea, it is, in effect, worthless. Once we vet and sell our ideas, we must implement them because some of the best ideas never yield benefits, as they are not implemented. This is what JMT does for us and this is what we are going to learn how to do this week. If some of the participants are engineers or engineer types, they may believe they know this already and all engineers should. What they do not know, which will be very helpful to them, is a standard method of applying this improvement method, which they and everyone in the company can use. They thus will have a common language to be used throughout the company so they will be able to gather more ideas and more easily implement their own ideas. In companies that have a formalized CI program, a weakness often is that the program does not include everyone. They may even have a standard process for working through opportunities to make improvements, but they often are so involved or so complicated that it is difficult to teach everyone to use them.

In companies with a formal CI department, the thinking is that the department is responsible for making the improvements, so the culture is such that ideas come from the top and go down, instead of coming from the floor and going up. If 100% of the employees are not involved, it is not really a true CI program. Furthermore, many good ideas are being wasted because they are never heard.

The summary to this point is

- We all want to improve the four metrics of quality, cost, delivery, personnel.
- One concept that will help us with that is continual improvement.
- The only way to get company-wide continual improvement is by having everyone know and use a standard improvement method, and JMT gives us the simplest workable method.

Similarly, JRT starts with the four metrics. In this program, the lead-in question has to do with the personnel metric. What must happen for the personnel metric to contribute in a positive way to the other three metrics? The answer is that there must be strong, positive personal relationships among all employees in order to foster the communication, cooperation, and teamwork needed. The question then is, how do we get strong positive personal relationships? The answer again lies in the five needs. This time, however, JRT is the last J program we discuss in the five needs.

The summary to this point is

- We all want to improve the four metrics of quality, cost, delivery, personnel.
- One concept that will help us with that is creating and maintaining strong, positive personal relationships among all employees.
- One way to create and maintain these relationships is to develop and practice appropriate skills such as those found in JRT.

Thus, the first hour of the first session tells the participants what each of the J programs is about, and how each program ties into and supports the main objectives of the company. They should now see that the J method they are about to learn serves as a foundation for what they do in the company.

Now we will continue with a discussion of each of the programs individually, since they differ after this introduction.

Job Instruction Training

Format

As a consultant, I always recommend having two sessions to maximize the use of my time for the client. One client selected to have only one group of 10 JIT participants because she wanted the remaining time spent in refining the skills of creating and instructing with JBSs. I recommend that in-house JIT trainers deliver only one JIT session a week. They will have other JIT responsibilities to attend to once the program gets started. In addition, several years ago I added a coaching component to the week's training. When I started delivering JIT, I delivered it the same way the person who showed me delivered it. I was never satisfied with the quality of the demonstrations after either he or I did the training. The format for the training we used was the same as that described in the original Trainer's Manual. That would be to deliver the four-step method on the first day and, at the end of the two hours, ask for one or two volunteers to demonstrate the method on the second day. The volunteer's demonstration on the second day would not be perfect, which would lead into improving the four-step method by using the JBS. The JBS would be introduced and discussed, and the second session would end. Between the end of the second day's training and the start of the third day's training, the volunteers would write their JBSs. When I recognized that the demonstrations could be improved, my first thought was to offer my help to the participants. A few accepted, but many wanted to do it by themselves so they would not have to have another scheduled time. The improvement in the quality of the demonstrations was obvious. The participants who had been coached were still not perfect, but they were consistently better than those who had not been coached. I would be with a participant for about an hour on a one-to-one basis. It was interesting to me that each participant seemed to hear different ideas in the first four hours of training. Although I was the only person delivering the material during those four hours, most people heard different ideas. Much of the hour of coaching can be spent just getting the participant to understand what a "job" is because most people do not think of breaking down a job as finely as is required for a JBS. When I offered my help at one client's facility, the coordinator said it would not be voluntary, and every participant would receive the one-hour coaching segment, and every person should schedule a time with me. A main reason all of my clients did not automatically seize on this opportunity is that many had the idea that the J programs require only 10 hours of training. That is true, but the one-hour golf lesson you take is backed up by many hours of practice if you wish the lesson to be worthwhile. It also occurred to me that delivering JIT as I was shown was not using the JIT method. If you really want people to accept and use JIT, you must use the JIT method when you are delivering JIT. I had done Steps 1, 2, and 4 but I had left out Step 3—Try out performance. Step 3 is what I do during the one-hour coaching session. The elements of Step 3 that I had been missing are one-to-one training, correcting a mistake as soon as it happens, and continuing until I know the participant knows. Although it does make for a long day on Tuesday, Wednesday, and Thursday,[9] I find it is well worth the effort because I know that the 20 people who receive the training actually know how to use the JIT method.

Previously, if a participant had a wrong idea about something, they would use that idea in their JBS and thus in their demonstration. This would only tend to solidify the incorrect idea in the participant's mind, making it that much harder to overcome with the correct idea. In addition, there often is not enough time to fully correct incorrect JBSs during the 2-hour sessions.

Developing Coaches

Because I am an external consultant, I am on-site only for the week of training. I do offer advice and guidance before and after the week for whatever the client requires, but that is done remotely via emails or telephone. Some clients would have my contact person become the TWI coordinator. This person would stay with me during the entire week of 40–50 hours. After the week of training with such a client, I knew the program would be successful. The coordinator had seen me coach employees in breaking down 20 jobs. By the end of the week, they were doing the coaching themselves, so I knew they could continue this after I had left. I then thought of the clients who had not assigned a coordinator to "shadow" me. How would their program fare? I decided that I must also develop at least one coach (and two coaches would be better) for every week of JIT. Once a person breaks down 15–20 jobs, they have a sufficient understanding to assist others. They are not fully experienced, but they are miles ahead of others in the group. The hours spent coaching the participants one-to-one in their JBSs can now be used to develop coaches. I ask for two of the 20 participants to shadow me during the week. At the end of the week, they will become coaches who can assist the other 18 participants in using the JIT Program. Refer to Figure 7.5 to see a typical week's schedule. Note that the schedule in

Coach development schedule 5-day week

Hours	Monday	Tuesday	Wednesday	Thursday	Friday
			TW1 – job instruction training		
1 and 2	Institute Conductor (IC) delivers 2-hour training (demonstrations Wed., Thurs., Fri.) 1st Coach candidate (1st CC) participates with nine others				
3, 4, 5	Discussion of JIT implementation plan; what areas to start with; what problems have been identified	IC coaches (3) participants while 1st and 2nd CC observe	IC coaches (3) participants with assistance from 1st CC; 2nd CC observes	1st CC coaches (3) participants while IC and 2nd CC observe	Discussion of implementation plan; how it will be executed in interim weeks
6			Lunch (optional)		
7 and 8	Institute Conductor (IC) delivers 2-hour training 2nd Coach Candidate (2nd CC) participates with nine others;				
9, 10, 11	Discussion of JIT implementation plan; what areas to start with; what problems have been identified	IC coaches (3) participants while 1st and 2nd CC observe	IC coaches (3) participants with assistance from 2nd CC; 1st CC observes	2nd CC coaches (3) participants while IC and 1st CC observe	Discussion of implementation plan; how it will be executed in interim weeks

Time required:

    1st Coach candidate (1st CC) – 28 hours   It is recommended but not required that both coach candidates
    2nd Coach candidate (2nd CC) – 28 hours   observe the other group. This would add 10 hours to their schedule.
    Participants – 11 hours

Figure 7.5   Coach development schedule.

Figure 7.5 does not specify hours. You must select the hours to best accommodate the participants, not the trainer. The sessions are best when they are held at the start of a shift because that is when people are most receptive, although it is recognized that this is not always possible.

Developing JIT Trainers

Developing coaches to assist the other participants as part of the week's training was a new idea born of necessity, but developing trainers existed from the origination of JIT. In the 1940s a small company did not need to have an in-house trainer because they could use the free services of the local district representative when needed. Larger companies would require the continued use of a JIT trainer, and so an employee must be developed for that position. The TWI Service would hold JIT Institutes led by an Institute Conductor (now often called a Master Trainer), where about ten employees from different companies would learn how to deliver the JIT program.[10] The week would consist of delivering the material in the JIT Trainer's Manual to the other trainer candidates. JBSs would be included in this week. At the end of this week, each trainer candidate would return to his company and prepare for a week of training. When preparations were complete, the Institute Conductor would be called to observe the first week of training. At the end of the training, the Institute Conductor would give a written review with a pass or fail grade. If the Institute Conductor thought he did well, the person would be given a JIT trainer's certificate. If the Institute Conductor felt that improvements were required, he would note the required improvements. The trainer candidate would practice afterwards and recall the Institute Conductor for another week's training when he was ready to try again. Note that the U.S. government covered all of the Institute Conductor's time and expenses. Today, the situation is slightly different because the company must pay for all training expenses unless they are fortunate enough to obtain a grant of some kind.

My first "train the trainer" sessions were conducted as I had been shown, which closely matched[11] what the TWI Service had done before, and which is the method most consultants use to develop JIT trainers. It took me two or three such sessions to recognize the deficiencies in such a method. The most obvious deficiency is that at the end of the week, the trainer candidates are completely overwhelmed and exhausted. They have spent 40+ hours repeating the same material over and over because each trainer candidate must have a chance to demonstrate his or her skill. Another deficiency is that each trainer candidate had only one chance to deliver the material, but there is not enough time for everyone to deliver the complete four hours of Sessions 1 and 2. Four hours times 10 people is 40 hours, which leaves little time for anything else. Reducing the recitation time to 2 hours would leave 20 hours for feedback, JBS practice, discussion of strategy and so forth. Having only 2 hours to demonstrate that you can conduct a JIT session is something an experienced trainer could do, but it should not be expected of a trainer candidate.[12] Another error in this method is that the trainer candidates are given a certificate at the end of the week. Their bosses spent money for a week of training, plus the necessary travel expenses, so they expect the person will be a trainer at the end of the week. At this point, however, they have not actually delivered the material to real participants and they have not had the Institute Conductor observe them doing so. The result of this is that the new "trainers" often lack the confidence to conduct their first session, and continue rehearsing until they gain that confidence. The problem is that the longer they wait, the more their confidence can ebb.

I felt that there had to be a better way, so I thought about not only what was required of a JIT trainer, but also about the JIT method itself. A JIT trainer must know the JIT material well enough so s/he can deliver it so the participants, get the participants to understand *how* to use it, and *want* to use it. The Trainer's Manual is a guide during the 10 hours of training, much like a JBS, so it should not be read but must be referred to. The trainer must also know how to facilitate the sessions. That is, the trainer must know how to answer questions from participants who have little or no knowledge of TWI and also be able to control the discussions. In addition, the trainer must know how to break down jobs and deliver instruction using the JIT method since this is what s/he will be teaching. The scheme I now use to develop trainers starts with the coach, who will now be referred to as a trainer candidate. This is a person who has spent 40+ hours watching the JIT delivery and breaking down jobs. I then have this trainer candidate assist his coworkers in starting to use JIT. They will break down jobs, get consensus, instruct some employees, measure gains, and so forth. I then have him or her study the JIT Trainer's Manual[13] and practice the delivery of the material. Before they start their practice deliveries, they should write a JBS for each day's lesson. Each day has Advancing Steps and Key Points because it is a job like any other. When you break down a job, you will know that job better than you ever have before and the same holds true for JIT. Thus, writing a JBS for each of the five days gets the trainer candidate to know the material really well so that they can deliver it better. In writing such a JBS, the trainer candidate must ask what really advances the training and what are the Key Points to be mentioned. Asking these questions embeds the knowledge of the sessions into the trainer candidate's mind, making him better prepared to deliver the material. The trainer candidate should practice delivering the material in front of a mirror, and then perhaps to his pet dog or cat, and then to a small group of people. A fellow coach or another participant would be a good choice. Ask for feedback so that you can improve.[14] Once the trainer candidate has enough confidence to deliver the material to a live group, we hold another week of JIT. The time between this week and the first week of training should not be less than four weeks because even a person who is 100% dedicated to the effort will take at least that long to prepare for the second week. During this second week I would deliver the morning session while the trainer candidate observes. Then the trainer candidate would review the material with me before the afternoon session covering any final concerns. Then, the trainer candidate would deliver the afternoon session while I observe. The reason this works well is that the trainer candidate is fully prepared and knows that s/he will be delivering JIT to his or her peers, so they want to make sure they do a good job. They have the manual to follow and I will be observing so they cannot make any significant error. It's much like running alongside your child when he's learning to ride a bicycle. He's doing it all by himself, but you are there to catch him if he starts to fall. At the end of each day, I give verbal and written feedback. At the end of the week, the trainer candidate has delivered his or her first week of JIT and can now be considered to be a JIT trainer. This is an almost mistake-proof method because (so far) it has always resulted in trainer candidates becoming competent trainers. I use the word "almost" because the sample size is not large enough to be 100% conclusive. There were two instances where the trainer candidates were not prepared. When we got to the point where they were to review the material with me before they delivered it to the afternoon group in the second week, they realized that they were not prepared. Consequently, I delivered the afternoon session and the trainer candidates remained functioning as coaches. Just as with any skill, performing a job and training someone how to do that job require two different skills.

Training: Delivering the Sessions

We will now discuss the remainder of each session in the week. The first two-hour session should be delivered just as the original JIT Trainer's Manual described with the addition of the four metrics. The second session as described in the original manual is poorly done and does not follow the JIT method, so some changes should be made. Sessions in Days 3–5 consist mainly of demonstrations plus the use of a Training Timetable.

Let's take a look at each session more closely.

Day 1

We will continue from the point where we leave the Five Needs of a Supervisor and begin the JIT discussion. We want to impress upon the participants the importance of training and begin to do that by noting that most of the problems we encounter would not occur if people had been trained properly. People do not want to make mistakes, but they do because they are often learning on the job by trial and error. Trial and error by definition means mistakes are made. It is only when those mistakes are large or injurious that they are noticed, but they do happen frequently. As a result, everyone requires training because jobs, criteria, policies, and so on, are constantly changing. Because of this, being able to instruct is one of the most important skills any person can possess. When we think of instruction as transferring knowledge, we realize that it is something we do every day. Knowing this makes the skill even more important and coveted.

Next we give and discuss the definition of Job Instruction

Job Instruction is THE way to get A person to *Quickly remember* how to do *A job correctly, safely and conscientiously.*

The way = There are many ways to instruct someone, but if you use any other way, it will take you longer or it will not work as well.
A PERSON = JIT is always one-to-one since that gives the best results; for example, a classroom teacher versus a tutor.
Quickly remember = the objective of JIT is to get a person to production speed as quickly as that person is capable.
A job = the amount of actions a person can understand and remember during one instruction session.
Correctly = the way she or he is showed standard work.
Safely = we will combine safety training with production training so people think about them the same way and always do both.
Conscientiously = we want people to care about their jobs because that results in overall quality. People may be doing their job correctly, but if they do not care about the end product, quality may suffer. We get people to care by telling them why their job is important.

Everything up to this point has been said to enable the participants to see how instruction fits in with the overall company objectives and why it is so important. We now will discuss how instruction is conducted at this specific facility.

There are four main techniques people use when they instruct someone:

- Showing
- Telling
- Questions and Answers
- Documentation

They all must be done correctly or the instruction will be poorly delivered. Documentation is a broad category and can include work instructions, schematics, diagrams, checklists, drawings, models, and any other training aids. The trainer wants to show the participants the errors people make in using these techniques, but for best use of time s/he will demonstrate the two we use most often—telling and showing. The best way to do this is with a sample demonstration job. However, if the trainer demonstrates instruction by combining showing and telling, which is how we usually instruct,[15] it will be difficult to separate the errors of each one. To make it easier to see the errors often made, the trainer separates the techniques. The first instruction demonstration will be done by only telling someone how to do the job. Then the instruction will be repeated, but this time the trainer will only show the volunteer. In each case the other participants should observe and critique the instruction. The trainer now recites how to tie the Fire Underwriter's Knot. Note that this must be done completely from memory, and with confidence, because the object is to display how an expert would tell someone how to do something. If the trainer reads the text, s/he is not simulating an expert and the volunteer might ask to see the paper. I have never had anyone be able to tie this knot after merely telling them how to do it. The trainer's objective now is to get the participants to tell him what went wrong. The following are some reasons the volunteer could not tie the knot just by listening.

- Many people are not auditory learners.
- We often give too much information at one time.
- We often give irrelevant information.
- We may use technical terms.
- Even a simple operation is difficult to describe in words.

- Few of us use the exact words required.
- Actions seem more complicated when we are just listening.
- It is difficult to know how much to tell and to determine if it's understood.

The trainer gets agreement that instruction often contains these flaws and the participants have even committed these errors. The question then is, "Who is responsible for the participant not being able to tie the knot?" The usual response very quickly is that it was the trainer's fault.[16] The trainer then writes the JIT motto on the board: "If the person hasn't learned, the instructor hasn't taught." The trainer must get everyone to agree with this and that usually happens. In order to check for anyone who has unstated reservations, the trainer asks if there are any exceptions to this. The exceptions usually stated are that either the person doesn't want to learn or doesn't have the ability to learn. It is important that everyone fully agree with the motto because we want to avoid an instructor saying someone cannot be trained.

The trainer states that there are three reasons jobs do not get done correctly:

1. The person can't do an action: Having to do with aptitude
2. The person won't do an action: Having to do with attitude, or
3. The person doesn't know how to do an action: Having to do with prerequisite knowledge

Examples of people who can't do a job because they lack the aptitude could be a colorblind person selecting parts based on color, a short person stacking tall shelves, or a person with poor eyesight seeing small details. If a person won't do the job, it may be because he is going through a personal crisis, he doesn't want to get dirty, or he thinks he should be doing something he considers more important. When we say a person doesn't know how to do a job, we are referring to any prerequisite knowledge required in order to do the job. If you are instructing someone how to inspect a part and that job includes the use of an inspection instrument such as a Vernier, or a counting scale, they must know how to use the required instrument because that is only part of the job. If they do not know how to use the instrument, they will have difficulty performing the job. The point of this is that the good instructor will check the learner to make sure s/he has the aptitude, the attitude, and the prerequisite knowledge for the job. If any of these three requirements is missing, the instruction should not begin until the situation is corrected. Being aware of any of these roadblocks does not mean that the instructor must correct them. It merely means that if the instruction proceeds, it most probably will not end well. Some deficiencies will be easy to address, such as providing a stool or ladder to access something out of reach or providing a magnifying glass for someone with poor eyesight. Some will be more difficult, such as when a person does not want to be trained because he thinks he is in the wrong job and wants to transfer. The instructor may not be able to correct this, but he should understand the situation and take appropriate action. If the situation is ignored, the result of the instruction will be less than adequate. The instructor must recognize that if all three requirements are present, it is up to the instructor to make sure the learner knows how to do the job.

The demonstration is repeated for "showing" someone. In this case, it is possible that a person can duplicate tying the knot just by watching. They have had the benefit of hearing the directions, seeing what the wire looks like, and even seeing it being tied once. I have not kept accurate records on this but my estimate is that about 1 person in 200 can tie the knot just by copying the trainer. If this happens, check to see if it has been tied correctly. The knot itself may be correct, but check to see that the loose ends are long enough and are the same length. In any event you can ask the volunteer if it is important that the ends are even or if there is a specified length for the loose ends. Use some examples of how copying motions can get us into trouble if we do not know why we are doing those motions. One example is in using the 5S Procedure. The second pillar of 5S is to "set in order." One way this is done is by outlining. When tools are hung on a board, it's helpful to outline them so it is easy to see what tool is missing from the board. Naturally, this can be used for any object such as refuse containers. In this case, tape would be put on the floor outlining where the container should be. Walking through an area that had been "5-s'd," I noticed that a table that was bolted to the floor was outlined and a vise bolted to the table was outlined. Naturally, these items would not be easily moved and must be used in place, so outlining was not necessary and was a waste of time and tape. The person who did it did not know the purpose for outlining objects. He just knew that some objects had to be outlined.

The faulty training is summarized as follows:

- The inadequate instruction displayed in the demonstrations are universal weaknesses and thus are in most instruction

- People can learn their jobs with enough telling, showing, and time, or through trial and error, but
- These are not sure and dependable ways to instruct
- There is a sure and dependable way that works every time if it is applied correctly
- The trainee learns what you want him to know as quickly as he is capable
- This method began about 100 years ago and has been used in offices and shops around the world for over 75 years
- If the person has the aptitude, the attitude and the prerequisite knowledge, it is up to the instructor to deliver the instruction so s/he can do the job

The trainer then demonstrates the JIT method by using a third volunteer. This is an important demonstration because it is exactly how the participants should be delivering instruction. Refer to Appendix 4 for a script that conveys how the instruction should be conducted and notes that can be included in the Participant's Manual. The trainer introduces the demonstration by saying that this is the method that results in good instruction. He asks the participants to listen closely because he wants them to understand the method of instruction and not just how to tie the knot. After the demonstration, he will ask them what he has done.

At the conclusion of the demonstration, the trainer asks the group what he did. The trainer attempts to get the actions in chronological order, but if they are not stated in that order, he writes them on the whiteboard in chronological order, leaving spaces when actions are not given in chronological order. He also mentally separates the whiteboard into quadrants and puts the actions that are done in each step into an appropriate quadrant. The quadrants and the actions are not numbered but just listed. The participants just described what they saw and, since what they saw worked, they are more likely to adopt the "new" method. Once all the actions have been listed, the JIT pocket cards are distributed and the participants are told that they just described to the trainer the JIT method. The trainer returns to the whiteboard and labels each step as he explains it.

Step 1

1. Put the person at ease: People will learn better when they are not anxious
2. Find out the person's knowledge—to gauge ability and prerequisite knowledge to see if other training is required

3. Get the person interested: By telling the importance—to support quality
4. Make sure the person can see what you're doing: Showing is important

Step 2

Deliver the information in "chunks"

Step 3

1. See the person perform the job
2. Have the person explain the job to you (Advancing Steps, Key Points and Reasons.) so that you know s/he knows

Step 4

1. Follow up: To make sure there are no questions and the job is being done correctly
2. Taper off the follow-up: To let the person know s/he has the job
3. Refer a resource for questions: To make sure the person gets the correct answers

To be doubly sure that the participants have accepted the new method, you can list the mistakes made in the "showing and telling" demonstrations and show how they are addressed by using the JIT method (Figure 7.6).

The only task that remains is to have a participant offer to demonstrate a job in the next day's session using the four-step method. If a demonstration matrix has been made, then there would be little discussion. The demonstrations should reflect what was done in the Fire Underwriter's Knot demonstration of the JIT method. The trainer should confirm with the volunteer that the volunteer will be present the next day. If there is any question at all, a back-up volunteer should be chosen. If all the participants have been briefed as part of the preparation, then this would have already been done. The trainer should review with the volunteer after the close of the session what will be done in the next session and how to prepare for it. The volunteer should have enough parts to perform the job several times and should use the four-step method. If JIT has been used in the company for some time, the trainer should ask the volunteer if s/he knows anything about JBSs. If the person does, the person should be told not to use one for this job and to instruct as would be done without the JBS.

Job instruction training techniques for good instruction
Enabled through the JIT method

#	Where used	Technique
1	Step 1	Check physical ability—confirm capability
2	Step 1	Check desire—confirm wants to do job
3	Step 1	Check pre-requisite knowledge—confirms qualifications
4	JBS	Give only relevant information—only what is necessary to do the job
5	JBS	Be concise but understandable—facilitates understanding and retention
6	JBS	Include all details—for required Q, P, S, C
7	Steps 3 and 4	Repeat as needed—retention
8	JBS	Give Information in order—facilitates understanding and retention
9	JBS	Use known terminology—avoids confusion and intimidation
10	JBS	Use the same words—facilitates understanding and retention
11	JBS and Step 2	Deliver in 'chunks'—facilitates understanding and retention
12	JBS	Explain why—increases interest and understanding
13	Steps 3 and 4	Check real understanding—no assumptions = fewer errors once on job

Figure 7.6   Techniques for proper instruction.

Day 2

The original manual does not explain the concepts of Advancing Steps and Key Points well enough for most people (including me) to understand them. When I was first shown JIT, I was told the definitions of both an Important Step and a Key Point. When I questioned one or the other, the definitions were repeated. Table 7.1 shows the timeline for Day 2 from the original JIT manual published in 1944.

The purpose of the volunteer's demonstrations is to show the necessity of the JBS. A JBS is paperwork and people usually do not like paperwork (unless it's their paycheck). The demonstration of the four-step method *without* a JBS shows that the four-step method is necessary, but it is not sufficient for the best training we can give. I believe that one such demonstration is sufficient to make the point. Even one such demonstration must be dealt with carefully because some participants may get offended that they were used as an example of what not to do. The volunteer should be praised for the job he did because the instruction was better than most people receive.

Table 7.1 Original Timeline JIT Day 2 (1944)

#	Time (minutes)	Cumulative Time	Activity
1	5	5 minutes	Open the session
2	15	20 minutes	First volunteer's demonstration
3	5	25 minutes	Comment of demonstration
4	20	45 minutes	Second volunteer's demonstration plus comments
5	5	50 minutes	Summary
6	15	65 minutes	First two "Get Ready" points, break down the Fire Underwriter's Knot, explanation of the JBS, Important Steps, Key Points and examples of Key Points
7	35	1 hour 40 minutes	Break down both volunteer's jobs, hand out 6 sample JBSs, discussion of two of them that depict cascading, summarize
8	5	1 hour 45 minutes	"Get Ready" points 3 and 4
9	15	2 hours	Summary, finalize jobs for each of the remaining participants.

The area I question the most is item 6, which gives just 15 minutes for not only breaking down the Fire Underwriter's Knot, but also explaining Important Steps and Key Points. From experience, I see that most people have difficulty with these concepts. The original manual not only does not explain the difference between the two actions, but also allows only a few minutes to explain them. Because of a lack of time, only the definitions are given, which, I believe, is why most people have a difficulty understanding them. This is a different way of thinking for most people and so they should really understand the concepts first before they start using them.

Table 7.2 shows my modified timeline for JIT:

The following are the contents of each section.

Item 1 (15 minutes: Open and Review)

The group has been away for 22 hours, so I want to bring them back and confirm as much as possible of what they know with a brief review. Most likely everyone did not absorb (or even hear) everything from the previous

Table 7.2   Contemporary Timeline JIT Day 2 (2015)

#	Time (minutes)	Cumulative Time	Activity
1	15	15 minutes	Open and review
2	20	35 minutes	Volunteer's demonstration
3	5	40 minutes	Comments of demonstration
4	15	55 minutes	JBS intro; concepts of AS and KP
5	40	1 hour 35 minutes	Breakdown Fire Underwriter's Knot
6	20	1 hour 55 minutes	Breakdown volunteer's demonstration
7	5	2 hours	Summary, close

session, so this is an opportunity to make it stick. Trainers are not perfect and, although they will be following the manual, something could be missed or misinterpreted, so this is also an opportunity to emphasize points. Day 2 rests on Day 1, so it is important that everyone have the same understanding before going ahead. This is why someone who misses the first day should wait until the next JIT week session.

Items 2 (20 minutes: Volunteer's Demonstration) and 3 (5 minutes: Comments)

We are attempting to show that, although the four-step method results in very good training, it can be elevated to an even higher level. The trainer must be very observant and find any faults that would be corrected by using the JBS correctly. If the volunteer does bring a JBS, much of the leverage of getting people to use a JBS is gone. If the volunteer had a good JBS and delivered it well, the techniques he used can be mentioned. In any case, the trainer could ask the volunteer why he used a JBS. That should create a lead-in to the next item. Before going on to the JBS, list the techniques that we want to use to improve our instruction. Refer to Figure 7.6. Together we have identified problems in instruction and we have a list of WHAT it is we should do. Now we will learn HOW we will do it.

Item 4 (15 minutes: AS and KP Concepts; JBS)

We start with the four "get ready" points. The timetable will be discussed tomorrow and the JBS will be discussed now. Since the other two involve only a little discussion, I handle those at this time. There are three main preparation steps for using a JBS:

1. Identify details
2. Organize the information
3. Rehearse delivery

Rehearsing delivery will take place once we have the JBS written and approved. The first action we must take is to collect all the details in the job. When people are asked about a job they know well, they often omit details that have become 'second nature' to them. Since they are not mentioned, they cannot be transferred to the learner. Thus we must find out what those details are so we can include them in the instruction. The trainer should use examples to demonstrate to the participants that they are not aware of many details around them. Examples are the number of eyelets in a shoe, or the number of buttons in a shirt. A favorite of mine is a warning on a common household product, which says, "Keep out of reach of children under 6 years of age. Do not swallow. If more is used than for intended purpose, get medical help or call poison control immediately." Most people do not know that the warning is on tubes of fluoride toothpaste. Although, in Europe you will have to find another example.

Once the participants understand that we must find all these details, we must show them that they must be organized also. To do this, I show them two arrays of numbers. (Refer to Figure 7.7.) I ask them to tell me how many numbers there are in the disorder array, what the numbers are and to reproduce the pattern. They are given about 7 seconds. Someone with a photographic mind could tell you the answers, but most people have difficulty just determining how many numbers are there. I next show them the ordered array and everyone gets it. Order does make a difference in understanding and so we must use that when we deliver our training.

I next introduce the concepts of Advancing Steps and Key Points, as described in Chapter 3. The main thoughts the participants should get out of this item is that if an Advancing Step is not done, the job stops. If a Key Point is not done, Quality, Safety, Productivity or Cost *may or may not* be affected and the job *may or may* not stop. However, we want the learner to ALWAYS perform ALL Advancing Steps and Key Points. If we do include an action as a Key Point, we must validate that Key Point by giving a Reason. If we cannot give a Reason, we have no grounds for the Key Point. Thus, there are no

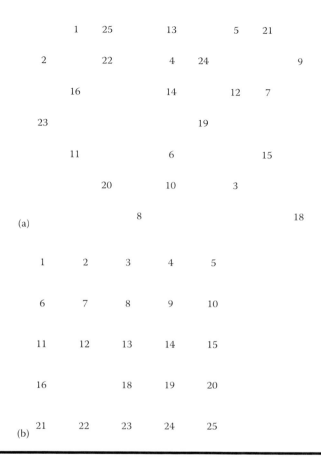

Figure 7.7 (a) Number array: Disordered, (b) Number array: Ordered.

shortcuts because the learner must do all Advancing Steps and Key Points in order to do the job correctly and as efficiently as possible on a continuous basis. Nothing should be added since it would only be a waste of time or effort, and would not improve the outcome in any way. Anything taken away would degenerate the instruction and the job. This is the best way we know to do the job at this time, but we are always looking for improvements. The next question is, "How do we present this information to the learner?"

Item 5 (40 minutes: Breakdown Knot)

The answer, of course, is by using a JBS during the instruction. A simple job that they are now familiar with is the Fire Underwriter's Knot, so that can be used as the first example. The original manual has the trainer talk to himself as he is breaking down the Fire Underwriter's Knot. That does make it go faster, but there are two things wrong with that method. First, we tell the participants that when they write a JBS, they must watch an expert do the job while they perform the analysis. One should not write a JBS by doing the job oneself because one does not have the perspective to see all the Key Points. Thus, we need an expert at the Fire Underwriter's Knot whom we can watch, so we can analyze the job and write the JBS. Since the trainer is not the only one in the room who knows how to tie the knot, the participant who was the volunteer on Day 1 (for the JIT method) becomes the expert. The second reason for using a participant is that the trainer has more opportunities for natural learning points. When the volunteer lays the wire on his hand to measure six inches, the trainer can ask what he is doing and if it advances the work. Often the participant will say that it does advance the work because he knows it is necessary to have a good knot. This facilitates the discussion of a Key Point not stopping the work but being required for, in this case, quality. Once we use the method to obtain all the Advancing Steps, we look for Key Points. Step 4 in the knot JBS does not have any Key Points and that offers a great learning opportunity. The following is the line of thinking that can be introduced here to get people to understand Key Points.

- Every step does not need a Key Point
- Every Key Point needs a Reason
- If a Key Point does not have a Reason, the Key Point does not exist
- An Advancing Step is WHAT we do. A Key Point is HOW we do it
- Therefore, if a Step has a Key Point, I must want you to do this Step a certain way since I am telling you HOW I want it done
- The way I am telling you to do the Steps affects Quality, Safety, Productivity or Cost
- If I do not tell you a Key Point, any way you do the Step will not affect Quality, Safety, Productivity or Cost, so you can do it any way you like
- This means that Key Points represent the variability in a job
- This means that Key Points represent the actions in a job where someone can make a mistake. Whenever a mistake has been made, it is because a Key Point was not done
- Key Points give us a measure of the chance of a job being done correctly. Many Key Points means that there is a high chance of a mistake being made. We thus want to reduce the number of Key Points in a job

- The best way to eliminate a Key Point is to change the process by mistake-proofing
- If a JBS does not have any Key Points, whenever the job is completed it would be correct

Much discussion after this may not be productive because there is a considerable amount of information here and the participants are usually mentally tired. The trainer must use his judgment, but the next topic also offers opportunities to embed the concepts of Advancing Steps and Key Points into the participants' minds.

Item 6 (20 minutes: Breakdown Volunteer's Demonstration)

This is done with the same technique that was used for the Fire Underwriter's Knot, and as described in the "Creating a JBS" part of Appendix 5.

Item 7 (5 minutes: Summary and Close)

The summary is that the JIT method consists of both the "Get Ready" points and the four-step method. Both are necessary to produce the best training. Review the "Techniques for Good Instruction Chart" and show where each technique is addressed in the JIT method. This should make the point that the poor instruction we used to do can be corrected by using the JIT method, and this is specifically how those corrections will be made.

Day 3

I supply each participant with a Participant's Manual. It contains a summary of the main points delivered and also gives space for the participants to take notes. The information is given in sentences with blanks, where key words give meaning to the concepts. Obviously if you do not know the key words, you do not know the concepts; so it is, in effect, a test of what the person really knows about JIT. A Reference Section includes other items such as "How to Create and Deliver a JBS," as is found in Appendix V. Since it is their manual, it can also serve as a reference for JIT after the training week. Day 3 begins with a review of this manual by making sure everyone has all the blanks filled in properly. Filling in the blanks leads to a discussion of the Training Timetable (Matrix) that was mentioned on Day 2 by saying we would talk about it on Day 3.

Training Matrix

A training matrix is a powerful tool to assist a supervisor at any level make sure the most important resource, the people, are fully prepared for production (refer to Appendix 1). We perform preventive maintenance to make sure our machines and equipment are always ready to perform and we spend a considerable amount of time making sure inventory is correct and up to date. It is surprising that many supervisors at all levels assume that all employees are fully trained to perform their jobs. One executive told me that the biggest lesson he obtained from implementing JIT was that he learned how much he assumed his employees knew. The Training Matrix gives us a quantified chart of the skill status of each individual so that we can really be ready for production. Contrary to 70 years ago, when most people did not use such a matrix, today, many people do use such a matrix or at least know of it. The original manual included each participant creating a Training Matrix for his or her own department or section. Today, I find that that this is not always a good use of time for several reasons. The only instance in which it would be a good use of time is when all the participants are supervisors (with direct reports) and management was not currently using such a matrix. The concept and use should always be discussed so that everyone knows of it, but it is not always necessary for everyone to create one. It is also helpful to have a discussion either during the session or with management about how the Training Matrix should be used. Trainers could update it, but it should be used by the line people in production, since resources for production are their responsibility. Just as the supervisor requests help from maintenance for repairing a machine, s/he should request help from trainers for completing training requirements.

The remainder of Day 3 is spent watching, critiquing, and discussing the three demonstrations that were coached yesterday after Day 2's training.

Day 4

Day 4 is spent entirely on watching and discussing the three demonstrations that were coached after Day 3's training. Often time remains for other activities that can solidify JIT concepts. I include a short movie that was given to me by a consultant who said it was used during the 1980s where he worked. It is now on YouTube and is titled "Problems in Supervision." The movie is a U.S. Government Office of Education Training Film that was made in 1944. A new version has been made, but the original offers more possible talking points. My main purpose for showing the film is to highlight the importance of the JBS. The supervisor in the film does not use a JBS, and yet the new worker seems to be trained very well. Neither the film nor the credits mention TWI or JIT, but the method is definitely closely related to JIT. The results in the movie seem to be good but that is because it is a movie and there is a script. In reality, although the training is better than what the supervisor did originally, it is not as good as it could be. The participants can see that without the use of a JBS, the operator in the movie really does not know what the Advancing Steps or Key Points are, which means she may omit them later on. Also, the training will probably vary from operator to operator. I find this movie to be useful and entertaining, but once the trainer has a good understanding of the time requirements, s/he should include anything that will make an impression on the participants.

Day 5

The main component of Day 5 is the three demonstrations. In addition to them, however, the trainer should take this opportunity to deliver or emphasize any points s/he judges will sell the program. The motto of JIT applies to the trainer during this week, so it is incumbent on him or her to find out what the participants have learned. Give them 5–10 minutes to take some notes on what they have learned this week and then go around the room and ask each participant what they learned about instruction this week. In addition to finding out how well the training went, this provides an opportunity to review the material one more time and to estimate the general acceptance of the program. This is also a good opportunity to explain what will happen next in the program. The worst path to take is to not have a follow-up plan. The company has spent time and money during this week, and if these participants do not employ this training fairly quickly, it will be a waste. If a month or two goes by without any action, much of what they knew will be lost. The participants should be told what will happen in the coming weeks so they know this is not a "program of the month." A feedback sheet should also be handed out because some people prefer writing comments than speaking in front of the group. The feedback should be anonymous, so people's true feelings and thoughts can be expressed. The feedback sheets will give the trainer information on what s/he should emphasize and may also include some suggestions to improve the training.

These sheets should be taken seriously but, aside from the suggestions, they really just give an idea of how the participants like the training and/or the trainer. This is why they are sometimes referred to as "smile sheets." No matter how positive these sheets are, they are only an indication of how the participants felt about the training and not an indication of the success of the training. The actual success of the training will be judged by how well the intended objectives are met.

Instruction: Beginning the Use of JIT

A week of JIT with double sessions is usually fairly tiring for the trainer, or trainer candidate and coaches, because they have spent a minimum of 40 hours immersed in JIT. But, on the other hand, they are usually very invigorated because they have seen the progress made with, and the possible benefits and opportunities of, JIT. The participants who have spent just the ten hours in the session in addition to the extra hour being coached for their demonstration will know JIT to the extent that they can break down a job and deliver it using the JIT method. However, they will not be particularly proficient at it and this is where the real work comes in. The plan that was made, whether it was a formal plan or an informal one, identified the part of the facility where JIT would be used first. That helped to determine who the first participants would be and what jobs would be considered first. Some of those jobs were probably used as demonstration jobs during the 5-day training, so some JBSs are already in a rough draft form.

The five days of training are usually done on Monday through Friday, so the actual use of JIT should start the following week. In some cases, the coordinator might want to take a week off before starting the actual use of JIT if this is starting out as a part-time job. A great deal of momentum can be lost if more than a week is taken before JIT is actually used, which is one reason it is very beneficial to have a full-time coordinator at the outset. It is a good idea at the end of the last day of training to tell the participants what will happen next. People are usually optimistic about JIT at the end of the training week, and that enthusiasm should not be wasted. The coordinator may not have all the answers to the implementation plan, but s/he should tell the participants as much as s/he knows. The following are points that can be covered as closing comments in the last minutes of Day 5, and that should be started on the following Monday.

- Who will write the JBSs?

    At this point, the only people in the organization who can write a JBS are the 20 employees who attended the 10-hour training. However, all of those people probably will not write JBSs because of their positions.
- When will they write them?

    When will it start and what will they stop doing to allow them time to write JBSs?
- What jobs will be done and in what order?

    The area has already been chosen and some of the JBSs have already been started, but additional information is still needed.
- What will the other participants do?

    Once employees have received TWI training, it's important to keep them included in the process. At a minimum they can function as a group that gives consensus of final draft JBSs.
- What is the process for getting consensus? For documentation?
- Who will do the instruction?

    Some companies have trainers in place already, while others use some of the participants in the first session. Note that in order to instruct using the JIT method, one must have attended the 10-hour JIT session.
- Where will the instruction take place?

    The ideal place to instruct an operator is at the workstation where the job is performed. However, there are some operations where instructing at the worksite would cause disruption in the output and so the training must be done off-line. There are many options between those two alternatives. For example, in one plant, an extra station was set up on the assembly line. The learner was instructed how to do the job, and any good product was added to the standard production. The output remained the same, but any excess product that the learner made was subtracted from the requirements of the experienced workers. When the learner's speed and quality reached that of the experienced workers, one person would be removed from the line. In another case, a conveyor was set up off-line so learners could be instructed and practice the job until they reached production speed. In some cases, such an off-the-line simulation cannot duplicate the production station completely, so the learner may not reach production speed before they are put on the production line. However, the "pre-training" significantly reduces the disruption caused by having a new person on the production line.[17]

- Who will collect, analyze, and publish data?

    The objective of this instruction was stated in the overall plan and would be something like reduction in the time people take to reach production speed, reduction in scrap or rework, reduction in discrepant product reports, and so on. Someone must set the baseline and recheck the measurements after the instruction has taken place. Finally, a report must be made so that everyone knows whether or not the objective was met.

Every organization will handle each of these points differently, but it is important that each one of them is covered. If the first week of JIT just ended, then the coordinator should get a possible date from the trainer candidate[18] when s/he will be ready for the second week of training, since those participants must be selected and notified. If this is the second week of JIT, then the coordinator must discuss with the new trainer when the next group of participants will be trained. It should be within a week or so, while the material is still "fresh" in the new trainer's mind; but if it's too soon, there may be more people to guide than the coordinator can handle.

Job Methods Training

Format

As with JIT, I recommend an outside trainer to deliver two sessions of JMT per week in order to give the client the most value for the cost of the training. An "in-house" trainer should deliver only one session per week. The purpose of JMT is to teach people how to vet, sell, and implement their ideas, and to do it in a standard way so everyone can easily understand it. The format used in the original manual is to develop the method by using a sample job. The job is first described and then demonstrated. Then the trainer describes what the person who made the improvements was thinking and what he did. The basic concept is good but the trainer does not involve the participants in any of the analysis. They are just told what the method is and what the results are. I have modified the format by first describing and demonstrating the job, and then asking the participants what improvements they would make. This not only gets the participants more involved in the training, which makes it more interesting, but also makes the concept of questioning everything more relevant. There are 16 changes that improved the job and the group of ten participants will usually identify most of them, but they rarely will see all of them.

Coaching

JMT does not contain any new concepts but just puts what people already know into a standard and easily repeatable method. As a result, the need for coaching of participants during the week is not as apparent. It is recommended to develop some participants to be able to coach others so that MBSs and improvements can be reviewed. Some people may need help or guidance in using the method, and it is always a good idea to discuss new ideas with someone to get a different perspective. It does take some discipline to continue to follow the method initially but that is the only way to maximize the number and quality of improvements. If coaching of participants is not done in the hours after the standard sessions, as is done with JIT, the trainer can spend some time with the coach candidates in additional drill. The idea of a coach should be considered when selecting the participants for each 10-hour session. Since the training takes only two hours, a section of a department or a product line could be in one session together. Thus the supervisor or group leader could take on the function of a coach when JMT is being used.

Developing JMT Trainers

Developing a person to be a JMT trainer would follow the same format as a JIT trainer. That is, one week of two sessions, where the trainer candidate participates in one and observes the other. This would be followed by several weeks of using and coaching JMT and that would be followed by a second week of JMT. Here the trainer candidate would observe the first session and deliver the second.

The Sessions

We will now discuss the sessions starting with Day 1 once we have finished discussing the five needs. We have established that continual improvement supports the four metrics, but the question is, "How do we accomplish continual improvement?"

Day 1

The underlying need for JMT is created because increasing productivity is important and, because it is so important, we should approach it methodically and not haphazardly. If we have a standard method for addressing problems (opportunities), we will not only solve the problem, but will also eliminate any wasteful actions.[19] We will have a standard method for mining and implementing ideas from all employees, and the best source of ideas is from those closest to the work.

The JMT Trainer's Manual states: "The Job Methods Program will help you produce greater quantities of quality products in less time by making the best use of the manpower, machines and materials now available."[20] Another way of saying this is that Job Methods will increase productivity and/or quality by eliminating waste. JMT eliminates waste by enabling people to see waste. The best way to explain the JMT program is to demonstrate an example. An important note is made that this example includes material handling, machine work, and handwork. In other words, every kind of work is included so someone cannot say that this does not apply to his job.

The job is described and then demonstrated. The trainer must be somewhat theatrical because he wants the participants to fully understand the job, and he will be in the training room and not a shop or the place where the work is actually done. There are many details in the "radar shield" job and the trainer must be careful to include all of them. Once the demonstration is complete, the trainer asks for suggested improvements and writes them on the flip chart. The trainer will act as the resource for the "company," since the participants will ask questions about the job that were not included in the stated description. The job consists of riveting together two sheets of thin, soft metal (brass and copper). The metal is thin and can easily be damaged so there is significant scrap. The inventory is located away from the workbench. The sheets are assembled with care because alignment is critical. Once assembled, they are stamped per the drawing with the word "TOP" in the lower right corner and put into a tote box. They are carried to a scale for weighing and then to shipping, where they are inspected and repacked into a shipping container, labeled, and weighed. The participants quickly see wasted time in walking to get the metal sheets and so their suggestions for improvements usually include moving the inventory closer. They may also ask why the shields are weighed before they go to shipping. The answer is that if the tote is overloaded, it will weigh more than is safe for the operators to carry. Thus, an inspector checks the weights periodically to see that the weights are within a safe range for carrying. A suggestion might be to use a conveyor or a cart to move the tote. When JMT is applied, the group can see that, although adding a conveyor or a cart does preclude the necessity of the scale, packing the shields at the workstation eliminates not only the scale, but also the double handling of the shields. This comes about only when all details are thoroughly questioned. The action that rarely gets questioned initially is the use of stamping "TOP" on the shields. Without JMT, people do not question that because it was said that the requirement to stamp was on the drawing. In JMT, we question the purpose of the stamp.

The trainer distributes the pocket cards and starts to analyze the job, with the participants using JMT. The MBS is used to record the information used in the job analysis, but filling out the form completely for this job takes longer than you have time for in this session. You want the participants to think through what they have to do, and telling them what to write in the blanks won't allow that to happen. Having them think through the process and fill in the blanks themselves (or even in groups) will take too long. Thus, you must compromise. The following method is a compromise that gets the participants involved in the analysis, but does not take an extraordinary length of time.

Step 1 is to break down the job by listing all the details. Explain that the more details you write, the more you can question. For people who have been through JIT, you should make sure they understand that there are many (JMT) details in one (JIT) step. Hand out a blank MBS and review the fields in the heading. Emphasize that the "Assisted by" field is important because you want to recognize everyone who helped you. Scratch out the word "proposed" and leave the word "present" readable because you will be recording the present method. Discuss the columns and the information that each should contain. Then begin the job and develop the first five details. Say, "What was the first action I took?" Get them to say, "Walk to the box of copper sheets." They may say, "Pick up some copper sheets," but you want a finer detail than that because the finer the detail, the more we can question. Include approximate distances where there are any and any notes they think might be appropriate. Write "1. Walk to box of copper sheets" on the whiteboard. Then continue asking what you did next and complete the first five details:

1. Walk to box of copper sheets   6 feet
2. Pick up 15–20 copper sheets

3. Walk to workbench   6 feet
4. Inspect and lay out   12 sheets
5. Walk to the box and return extra sheets   6 feet

Make the point that Bill Smith initially wrote the details as shown, but when he began to question them he recognized that Detail 4 is actually five details, so he re-wrote the MBS as follows:

4a: Inspect 12 sheets
4b: Lay out 12 sheets
4c: Walk to scrap bin (when required)
4d: Scrap copper sheet
4e: Walk back to bench

Have the participants continue to fill in details for about 10 minutes. This can be done individually, in pairs or small groups.

After about 10 minutes, hand out the completed MBS (Figure 7.8). We now proceed to Step 2—Question Every Detail. The pocket card lists the six questions we will be asking and the note states that we will question everything we can think of—nothing will be considered "off limits." This discussion should include some very important points. First, we want to develop a questioning attitude because answers to questions often lead to other ideas. However, we must be sensitive to the people of whom we are asking questions so that they know we are not questioning their ability. We only want to understand the process. Asking many questions can become annoying to people so we must be tactful and make sure they understand our objective. The questions must be asked in the correct order because if we ask "How" before we ask "Why" we will be wasting time if the detail can be eliminated. If we get a negative answer to "Why" or "What," we do not need to proceed further and question the next detail. The conversation for the first detail might be like this.

- Why do I walk to the box of copper sheets?
  – Because they are away from the workbench.
- Why are they away from the workbench?
  – Because that is where the material handler put them.
- Why did the material handler put them there?
  – When we ask the material handler, she says that is where she was told to put them.

Implementation ■ 211

Job methods breakdown sheet

Operation: Inspect, Assemble, Rivet, Stamp and pack   Product: Microwave shields   Department: Riveting and packing

Originator(s): Bill Brown   Assisted by: Jim Jones   Date: June 14, 1944

Present / ~~Proposed~~   (1)

List of all details for method — List every single thing that is done	Distance (feet)	Time (minutes)	Notes — Reminders, Tolerances, Etc.	Why?, what?	Where?, when?, who?	How?	Ideas — Write them down — Don't trust your memory
1 Walk to box of copper sheets	6		Placed 6 feet from bench by handler				
2 Pick up 15 to 20 copper sheets							
3 Walk to bench	6						
4(a) Inspect 12 sheets			Scratches and dents; scrap into bins				
4(b) Lay out 12 sheets							
4(c) Walk to scrap bin (when required)	3						
4(d) Scrap copper sheet							
4(e) Walk back to bench	3						
5 Walk to box and replace extra sheets	6						
6 Walk to box of brass sheets			Placed 3 feet from copper box by handler				
7 Pick up 15 to 20 brass sheets							
8 Walk to bench	6						
9(a) Inspect 12 brass sheets							
9(b) Lay out 12 brass sheets			One on top of each copper sheet				
9(c) Walk to scrap bin (when required)	3						
9(d) Scrap brass sheet							
9(e) Walk back to bench	3						
10 Walk to box and replace extra sheets	6						
11 Walk to bench	6						

Figure 7.8   MBS demo job present (Step 1).

Details	Distance	Time	Notes	Why? what?	Where, when, who	How	Ideas
12 Stack 12 sets near riveter							
13 Pick up one set with right hand							
14 Line up sheets and position in riveter			Line-up tolerance is 0.005"				
15 Rivet top left corner							
16 Slide sheets to the left and rivet top right corner							
17 Turn sheets and position them in riveter							
18 Rivet the bottom right corner							
19 Slide sheets to left and rivet bottom left corner							
20 Turn sheets around as you lay them on the bench							
21 Stamp "TOP" and stack them on work bench							
For Details 13 to 21 repeat the process 11 times							
22 Put 12 sets of shields into the tote box							
23 Carry the full box to the scale and weigh it	50						
24 Fill out a measuring slip and place it into the box			Approximately 75 lb gross				
25 Take the tote box to the packing area	100		By handler				
26 Unload the shields from the box			By packer				
27 Put 200 sets of shields into the packing box			Check inspection. Wood cases supplier by handler				
28 Cover and seal the box and write address slip on it							
29 Fill out a delivery slip							
30 Store box until it is delivered			Empty tote boxes returned by handler				

Figure 7.8 (Continued)   MBS demo job present (Step 1).

- Why was she told to put them there?
  - She has to put them some place. Where do you want her to put them?
- Can she put them on or near the workbench?
  - Yes

So on the MBS, we write an idea in the last column that the copper sheets can be moved closer to the workbench and this would eliminate the necessity of walking to get them. We also put a check in the "Why" column to remind us when we are devising the new method that this detail can be eliminated. It is important to note that we are analyzing actions and therefore we should be focusing on the verbs in the details. In Detail 1, we focused on the word "walk" and asked "Why do we have to walk to get the shields?" We have broken down the action of "getting copper sheets" into two actions: "walking" and "picking up sheets." The analysis shows us that we can eliminate walking but we can't eliminate picking up the shields. When we question Detail 2, we cannot eliminate picking them up unless we automate the process with a machine. JMT expressly states "best use of equipment now available." When we ask the remaining four questions we find that "the best place to pick them up" will be the workbench and not the shelf. Also, the question of "How" may have some possibilities. We think there must be a better way, but we do not know what it is just yet. Thus we check the boxes "Where, When, Who," and "How" to remind us that we will be changing the process on this detail. We add a note to remind us what we were thinking to do. Continue questioning the details up to 9(e) and then have the participants work by themselves for ten minutes or so. Again, this can be done individually, in pairs, or groups.

Hand out the completed present MBSs once you feel the participants have spent a sufficient amount of time on their own (Figure 7.9). Give the participants a chance to review the MBS and then begin Step 3. Step 2 and Step 3 have a symbiotic relationship in that Step 2 asks the questions and Step 3 results in the actions of eliminating, combining, rearranging, or simplifying. We go through each detail and, based on the notes we made on the MBS, we take one of those four actions. Give the participants a few minutes to arrive at a proposed method after giving them a blank MBS. Based on the time remaining, hand out the proposed method (Figure 7.10) and briefly discuss it. Make a point of showing where each of these actions has taken place. Eliminating walking eliminated Details 3, 4(c), 4(e), 5, 6, 8, 9(c), 9(e), 10, and 11. Details 22, 26, and 27 were combined and Details 2 and 7 were

Job methods breakdown sheet

Operation: Inspect, Assemble, Rivet, Stamp and pack Product: Microwave shields Department: Riveting and packing

Originator(s): Bill Brown Assisted By: Jim Jones Date: June 14, 1944

{ Present / ~~Proposed~~ }

List of all details for method — List every single thing that is done	Distance (feet)	Time (minutes)	Notes Reminders, Tolerances, Etc.	Why?, what?	Where?, when?, who?	How?	Ideas — Write them down — Don't trust your memory
1 Walk to box of copper sheets	6		Placed 6 feet from bench by handler	✓			No, if sheets nearer bench
2 Pick up 15 to 20 copper sheets					✓	✓	Close to riveter; better way
3 Walk to bench	6			✓			Same as #1
4(a) Inspect 12 sheets			Scratches and dents; scrap into bins		✓	✓	Just before assembly; better way
4(b) Lay out 12 sheets				✓			No, if sheets nearer bench
4(c) Walk to scrap bin (when required)	3			✓			No, if bin closer
4(d) Scrap copper sheet							
4(e) Walk back to bench	3			✓			No, if bin closer
5 Walk to box and replace extra sheets	6				✓		Same as #1
6 Walk to box of brass sheets			Placed 3 feet from copper box by handler	✓			Same as #1
7 Pick up 15 to 20 brass sheets					✓	✓	Same as #2
8 Walk to bench	6			✓			Same as #1
9(a) Inspect 12 brass sheets					✓	✓	Same as #4a
9(b) Lay out 12 brass sheets			One on top of each copper sheet	✓			Same as #4b
9(c) Walk to scrap bin (when required)	3			✓			No, if bin closer
9(d) Scrap brass sheet							
9(e) Walk back to bench	3			✓			No, if bin closer
10 Walk to box and replace extra sheets	6			✓			Same as #1
11 Walk to bench	6			✓			Same as #1

○ 2

Figure 7.9 MBS demo job present (Step 2).

Details	Distance	Time	Notes	Why?, What?	Where, When, Who	How?	Ideas
12 Stack 12 sets near riveter				✓			No, if no layout
13 Pick up one set with right hand						✓	Better way
14 Line up sheets and position in riveter			Line-up tolerance is 0.005"			✓	Better way
15 Rivet top left corner						✓	Better way
16 Slide sheets to the left and rivet top right corner						✓	Better way
17 Turn sheets and position them in riveter						✓	Better way
18 Rivet the bottom right corner						✓	Better way
19 Slide sheets to left and rivet bottom left corner						✓	Better way
20 Turn sheets around as you lay them on the bench						✓	Better way
21 Stamp "TOP" and stack them on work bench				✓			Find out
For Details 13 to 21 repeat the process 11 times							
22 Put 12 sets of shields into the tote box					✓		For details 22 to 30, counting and packing could be done anytime and anywhere after riveting
23 Carry the full box to the scale and weigh it	50			✓			
24 Fill out a measuring skip and place it into the box			Approximately 75 lb gross	✓			
25 Take the tote box to the packing area	100		By handler		✓		
26 Unload the shields from the box			By packer		✓		
27 Put 200 sets of shields into the packing box			Check inspection. Wood cases supplier by handler		✓		
28 Cover and seal the box and write address slip on it							
29 Fill out a delivery slip							
30 Store box until it is delivered			Empty tote boxes returned by handler				

Figure 7.9 (Continued) MBS demo job present (Step 2).

Changes summary

1. Sheets moved to workbench
2. Two riveters/station
3. Fixture to position riveters
4. Fixture to position sheets
5. Fixture to count assemblies
6. Jigs to hold and fan sheets
7. Scrap bins closer
8. Slots in bench for scrap
9. Shipping cases closer
10. Packing done by operator
11. Eliminate one scale
12. Eliminate stamp
13. Single piece flow—no batch of 12
14. Two operators only
15. Inexperienced operators instead of experienced/strong operators
16. Inspection no longer done @ packing

Figure 7.10 Changes in demo job.

rearranged. Actions were also simplified by using a fixture to align the two sheets and by having a magazine for dispensing the sheets. As mentioned, there is much detail in this job and it does not all have to be discussed, although the trainer should be prepared to discuss any of it. The major point is made when the changes that are included in the proposed method are compared with the suggested changes written on the flip chart (Figure 7.11). If no one suggested, change #12 (Eliminate stamp), the point will be well made that a thorough questioning can eliminate waste that we usually do not see. The quantifiable benefits should be reviewed so the participants can see the gains made by the improvements.

As with JIT, jobs can be pre-selected so that the participants can think about what they will be using for a demonstration. Emphasize that the improvement must be something that is new and not an improvement that has already been started. Sometimes participants will bring in ideas that that they are already working on. If that does happen, the trainer must question them on all the remaining parts of the job that the participants by-passed. This will be the procedure to be used for the demonstrations:

1. Describe the job.
2. Demonstrate the present method.
3. Explain the details and questions of the present method that led the participant to new ideas and the new method.
4. Demonstrate the proposed method.

Implementation ■ 217

Job methods breakdown sheet

Operation: Inspect, Assemble, Rivet, Pack   Product: Microwave shield   Department: Riveting and packing

Originator(s): Bill Brown   Assisted By: Jim Jones   Date: June 14, 1944

(3)

List of all details for method ⎰ Present ⎱ ⎱ Proposed ⎰ List every single thing that is done	Distance (feet)	Time (minutes)	Notes Reminders, Tolerances, Etc.	Why?, what?	Where?, when?, who?	How?	Ideas Write them down Don't trust your memory
1 Put pile of copper sheets in right jig			Boxes placed on table by handler				
2 Put pile of Brass sheets in left jig							
3 Pick up one copper sheet in right hand and one Brass sheet in left hand							
4 Inspect both sheets			Scratches and dents. Drop scrap through slots				
5 Assemble sheets and place into fixture			Fixture lines up sheets and locates rivet holes. Brass sheet on top				
6 Rivet the 2 bottom corners							
7 Remove, reverse, and place sheets into fixture							
8 Rivet the 2 top corners							
9(a) Place shield in front of fixture. Repeat steps 3-9 inclusive (19) times							
10 Put 20 Shields into shipping case—200/case			Cases placed by handler				
11 Carry full cases to packing department	150		By Handler with hand truck				
12 Close, weigh, and stencil cases			Check inspection by packer				
13 Write weight on delivery slip							
14 Set cases aside for shipment							

Figure 7.11   MBS demo job proposed (Step 3).

218 ■ *The TWI Facilitator's Guide: How to Use the TWI Programs Successfully*

5. Summarize the gains.
6. Describe the status of the proposal and any future actions.

The participants should come prepared to the next session with any tools and materials needed.

Day 2

It is important to review what was done on Day 1 and to be specific in going over the four-step method. Writing the method on a flip chart is useful, so that it can be referred to throughout the week (Figure 7.12).

It is now appropriate to introduce the proposal sheet. There are three main reasons for using a proposal sheet. The first is to help us evaluate the new method so that we can be sold on it. This sheet will tell us if the new idea is a good one. There is no need attempting to convince someone else of this idea if we cannot convince ourselves. And if we can convince ourselves, we should be able to convince others, if we have all the facts. A second reason for the proposal sheet is to sell the idea to others. The explanation will be brief and in writing, so people can read it at their own pace and on their schedule. We do not even have to be present when they review the new method. The third reason is that the proposal sheet

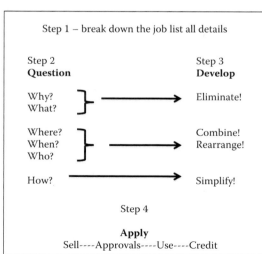

Figure 7.12 JMT summary.

documents the improvement so we have an automatic record of changes and improvements.

There are four parts to the proposal sheet:

1. Heading: Includes the individuals involved, the product(s), the location and the date.
2. Summary: A brief description of before and after the improvement, any gains and the status.
3. Results: Data that shows changes in various metrics such as quality, cost, productivity, and so on.
4. Content: This is a more complete description of what was done and the benefits achieved. Note that drawings, layout, sample documents, and so on can be attached to the proposal.

Aside from any attachments, the proposal should be kept to one single-sided page. The reverse side could be used for signoffs. All participants are required to write and hand in a proposal sheet. They can be tabulated and sent to the person who arranged and/or asked for the training. Refer to Figure 7.13 for the proposal sheet for the Radar Shield job.

Before the demonstrations begin, there is one more item to discuss: resistance and resentment. Resistance will be common because improvements include change. However, people do not dislike change; they only dislike the pain or disadvantages that come along with some changes. A well-written proposal sheet can reduce or eliminate much resistance because the advantages and disadvantages will be clearly written. Step 4—Apply—involves getting approvals before the improvement has been submitted, and that by itself should reduce the likelihood of much resistance. Resentment is caused when our questions convey the sense that we are attacking someone personally. That is not the objective and care should be taken to make sure that does not happen. Instead of asking a series of questions, it is often better to engage in a conversation that includes the questions. This will give you an opportunity to explain your true objective. There are many reasons for resentment and criticism, and each one must be treated individually and answered in a straightforward manner.

When a participant is demonstrating his use of JMT on a job, he should describe how he used JMT. It is not enough here to describe a new idea or method. We want to know that the participant knows how to use JMT so that the job can be made as effective as possible.

Job methods proposal

Submitted to: Ralph Horowitz	Date: April 15, 200x
Created by: Bill Brown	Dept./Cell:
Assisted by: Jim Jones	Riveting/Assembly
Product: Microwave shields	
Operations include: Inspect, assemble, rivet, pack	

Summary

We looked at the manufacture of the microwave shields and found that much waste could be eliminated by making a few changes. Excessive walking and handling did not add value to the product and could be eliminated by rearranging the workplace and using a fixture anda jig. The four riveters in the department would now be used by two unskilled workers instead of four skilled workers. The production increases 50%. The results are summarized below.

Results

Metric	Before	After	Difference
Quality			
Cost (fixture and jig)		$577	+$577
Production (department per day)	3200	4800	+1600
Safety			
Machine use (rivets/day/machine)	3200	4800	+1600
Reject rate (scrap)	15%	<2%	−13%
Number of operators	4 skilled	2 unskilled	−2
Distance (total travel by operators/day)	~1600 ft (4 operators)	0 ft (2 operators)	−1600 ft
Time			

Content

1. Operators no longer have to walk to the supply box if the material handler brings the sheets to the designated place on the workbench.
2. Placing two riveters inside a fixture and guiding on it to align the sheets will enable two holes to be riveted at once.
3. Having two jigs with sheets on top and placing them near the fixture will enable workers to use both hands to pick up the sheets.
4. Stamping 'TOP' operation was eliminated per ECO #567 – 23.
5. Emptying packing cases next to the bench will enable the operators to install the completed shields directly into the cases by number.
6. Having two slots on the bench will enable operators to drop defective sheets easily into the scrap bins.

Refer to the attached breakdown and layout sheets.
Attach present and proposed breakdown sheets, diagrams, and other related items as appropriate.

Figure 7.13 Proposal JMT demo job.

Proposal sheets are useful for at least two reasons:

1. **They help explain the improvement**. The sheet forces us to concisely describe the improvement by specifying facts. The improvement can be explained verbally with the sheet as guide, or it can be sent to someone without a verbal description. It helps us broadcast the improvement more easily.

2. **They document the improvement**. This documentation can be used as a measure of productivity increases or as a source of information when additional changes are considered. When people are unavailable, we can determine why changes were made. This record also helps to keep the process from experiencing slight changes that may become substantial over time.

The reverse side of this page contains a proposal sheet that was modified from the original training manual. Only the format has changed; the information is the same. The format was changed in an effort to make the form easier and faster to read.

Additional metrics may be added or the format may be revised further to better suit a particular organization. The intent is to give all the necessary information in an easy-to-read format so that many people can quickly see the benefit of the improvement.
Include

- How machines, manpower and materials are better used—be specific—use facts
- All improvements to quality, safety, housekeeping, design, workflow, etc.
- Names of people submitting and receiving the report
- All the names of people who worked on the improvement

Figure 7.13 (Continued) Proposal JMT demo job.

Job methods proposal

Submitted to: _____   Date: _____
Created by: _____   Dept./Cell: _____
Assisted by: _____
Product: _____
Operations include: _____

Summary

_____
_____
_____
_____
_____
_____
_____
_____

Results

Metric	Before	After	Difference
Quality			
Cost			
Production			
Safety			
Machine use			
Reject rate			
Number of operators			
Distance			
Time			

Content

_____
_____
_____
_____
_____
_____
_____
_____
_____

Attach present and proposed breakdown sheets, diagrams, and other related items as appropriate.

Figure 7.13 (Continued)   Proposal JMT demo job.

Before turning in your PROPOSAL, be SURE you have rechecked the New Method with the Job Methods 4-STEP PLAN.

STEP I – BREAK DOWN the job
1. List ALL the details of the job EXACTLY as done by the present method
2. Be sure details include all:
   Materials Handling
   Machine Work
   Hand Work

STEP II – QUESTION every detail. Use these types of questions.
1. WHY is it necessary?
2. WHAT is its purpose?
3. WHERE should it be done?
4. WHEN should it be done?
5. WHO is the best person to do it?
6. HOW should it be done "in the best way"

Also, QUESTION the -

MATERIALS
Can better, less expensive or less scarce materials be can the scrap from this job be used for another product?
Have the defects and scrap been reduced to a minimum?
Are the material specifications entirely clear and definite?

MACHINES
Is each operating at maximum capacity?
Is each in good operating condition?
Are they serviced regularly?
Is the machine best for this operation?
Should a special set-up man or the operator make all the set-ups?
Can use be made of the machine's or operator's "idle" time?

EQUIPMENT AND TOOLS
Are suitable equipment and tools available?
Have they been supplied to operators?
How about gauges, jogs and fixtures?
Have equipment, tools, been properly pre-positioned to permit effective work?

PRODUCT DESIGN
Could quality be improved by a change in design or specification?
Would a slight change in design save much time or materials?
Are tolerance and finish necessary?

LAYOUT
Is there a minimum of backtracking?
Are the number of handlings and the distances traveled at a minimum?
Is all available space being used?
Are aisles wide enough?

WORK-PLACE
Is everything in the proper work area?
Can gravity feed hoppers or drop delivery chutes be used?
Are both hands doing useful work?
Has all hand-holding been eliminated?

SAFETY
Is the method the safest as well as the easiest?
Does the operator understand all safety rules and precautions?
Has proper safety equipment been provided?
Remember, accidents cause WASTE of manpower, machines, and materials!

HOUSEKEEPING
Are working and storage areas clean and orderly?
Is "junk" taking up space that could be used for additional operators, machines, benches, and operations?
Do away with anything that is unnecessary.
Be sure necessary things are in proper places.
See that good shop housekeeping
reduces delays, waste, and accidents!

STEP III – DEVELOP the new method
1. ELIMINATE unnecessary details.
2. COMBINE details when practical.
3. REARRANGE for better sequence.
4. SIMPLIFY all necessary details.

STEP IV – APPLY the new method
1. SELL your PROPOSAL to your boss and operators.
2. Get final approval of all concerned on SAFETY, QUALITY, QUANTITY, COST.
3. Put the new method to work. Use it until a better way has been developed.
4. Give CREDIT where credit is due.

Figure 7.13 (Continued) Proposal JMT demo job.

Days 3–5

The main content of these days is the participants' demonstrations. If time is available, other content can be discussed such as the relationship between JMT and Value Stream Mapping, for instance. Day 5 should include a review, which can be based on the "What I Learned" exercise described in JIT.

Methods Improvement: Beginning the Use of JMT

JMT is used best when everyone in the organization has received the 10-hour program. If this was the first week of JMT, then the trainer candidate should be preparing for the second week, when s/he will become a trainer. If this is the second week of JMT, the new trainer should be selecting participants for the following weeks. One person (perhaps the TWI coordinator) should act as a clearinghouse and central repository for all JMT activity. The proposals and MBSs should be kept on file, and the savings and cost data from any improvements should be recorded and published.

There are two main ways to facilitate the use of JMT. The first is to just make the participants aware of the fact that they now have a tool for making improvements and that you want them to use it on any ideas they have. Each group of ten participants should have at least one person who knows JMT well enough that the other participants can go to him or her for assistance and guidance. This may or may not be the person's supervisor, although the supervisor should always be aware of the use of JMT. The employees must be given time for improvements, and how that is accomplished will vary with the organization. Some companies give their employees a block of time to work on any improvements they are thinking about while others take it on a case-by-case basis. One director told me that when he first took his position, he asked his supervisors how much time they spent making improvements. They all replied that they were too busy with production to spend any time at all making improvements. He told them to schedule about an hour a week or 15 minutes a day for making improvements. About six months later, he took the same poll and found that most of the supervisors were spending about five hours a week on improvements, with an accompanying increase in productivity.

The second way is to have a department, section of a department, or product line collectively reassess what they are doing. Everyone in this group must have gone through the JMT and everyone must be involved in the project. Everyone would assess his own job and some would also look at the overall flow through the department. Those who assessed the flow through the department would have to report to all department members. The first goal of this group would be to identify and quantify the intended improvements, and everyone would base their analyses on those objectives.

Job Relations Training

Format

JRT was developed for supervisors to deal successfully with personnel situations. The people who developed this program knew how a person should handle personnel issues, but they were not sure how to create a program that could be quickly taught in a standard way in a wide variety of companies. JIT had been very successful and JMT was becoming successful, and both were based on using a four-step method, so it was natural to create a four-step method for JRT. They took the scientific method, which they called the engineering method, and applied it to personnel problems. They did realize that there were other important factors that must be considered also and so they grouped these together and called them "Foundations for Good Relations." The concept was that the foundations would prevent personnel issues from arising, but if they did, they should be addressed with the four-step method. Because they were in an "emergency situation" they delivered JRT only to supervisors. The workforce was much more unionized then and when unions became aware of JRT and what it could do, they sought the training also. A separate program was created called Union Job Relations Training (UJRT). The main difference was one of terminology where "management" was actually union management (stewards, national representatives, union presidents, and so on). The unions coveted UJRT because when a union member had a grievance, his steward would use the four-step method to handle it. Fewer grievances were made and those made were usually won, saving the unions much money. The union officers then were the only operators who received the JRT program. Supervisors were the employees trained in JRT because they are the only employees who were paid to solve personnel issues. If an employee who has no reports (an individual contributor) has a personnel issue with another employee, he would seek help from his supervisor. Thus, many people feel that JRT need only be delivered to supervision. As a result, the focus of JRT seems to be the four-step method for solving personnel problems. However, the more important part of JRT is the collective foundations because, when used properly, they most often will eliminate the need to use the four-step method. If all employees were trained in JRT, far fewer personnel issues would develop and the necessity of the four-step method would be reduced. The thinking today is still one of delivering JRT to supervision only, but I believe that will change, albeit slowly.

Consequently, if only supervision is receiving JRT, there will be fewer eligible participants. All supervisors should receive JRT, since one's boss will become one's coach and that will extend all the way to the top executive at the site. JRT is different from JIT and JMT in that up to about 13 participants can be in one group. If there are about 14 or more, it would be better to have two groups, since sitting through 13 personnel problem analyses can be difficult. JRT is different also in that it is completely contained within the two-hour session. The method is given during the first session and then participants analyze a personnel problem with which they are familiar in subsequent sessions. No work is done outside of the sessions. An external consultant, therefore, might deliver only one 10-hour program in a week's time.

The material contained in JRT should be common sense to any truly successful supervisor. They should already know the foundations and facets of the four-step method. As a result, many managers, especially senior managers, believe they do not need to attend a JRT session. JRT, however, does benefit managers (including senior managers) in at least three ways. First, it gives them a standard method to use for what they have been doing intuitively. The result will be that their actions will be stronger and more consistent. Second, they should know what method every supervisor at the site is using to solve personnel issues, since they may have to critique some issues. Every supervisor should be a JRT coach to his or her direct reports. One responsibility of every supervisor is to develop the people who report to him or her, and this is a way to do that. A third reason for attending is that they will gain some insight about their peers, seeing how others handle situations.

A caution when selecting participants for JRT is that, in one group the management levels should all be roughly the same. The demonstrations will be personnel situations and the situation a vice president uses may include some people to whom a first-line supervisor reports. Thus, senior management, middle management, first-line supervision, and individual contributors should all be in separate groups. If it is necessary to mix these groups at all, the more senior management should use examples from experiences at another company.

Developing JRT Trainers

Although the material may not be new to many people who will be in a position to be a JRT trainer, the method is somewhat different for this application. The minimum requirements to become a JRT trainer should be to participate in one JRT session, to study and rehearse the delivery of the material, and then to deliver a 10-hour session while an Institute Conductor (Master Trainer) observes.

Let us now look at the main points in the five days of JRT, starting as we finished the five needs discussion. This section will be brief because the training follows the original manual since, aside from what has already been mentioned, no changes were deemed necessary.

Sessions

The main ideas are that every supervisor gets results through people. A supervisor obviously cannot do all the work and if s/he could, there would be no need for the people who report to them. The result a supervisor wants is a given output of a given quality. A supervisor's job is to get people to do:

- What should be done
- When it should be done
- The way it should be done because
- The people want to do it[21]

The only way to be able to do this is to have the cooperation and loyalty of people who report to the supervisor. That requires having strong, positive personal relationships with them. The way to develop strong, positive relationships is to use the Foundations for Good Relations.

In addition, the supervisor must treat everyone as an individual. That does not mean everyone should be treated the same way, but the same rules will apply to everyone. For example, everyone should be audited but some people may have to be audited more frequently than others. The exercise in the manual makes the point that people are more different than they are alike.

JRT defines a problem as anything that interferes with production.[22] This is important because some people think that a supervisor need deal only with the mechanics of the operation and social workers or psychologists should deal with the human side of the operation. A supervisor is not supposed to be a social worker or psychologist, but s/he must recognize when the assistance of one is needed. A supervisor's objective is to maximize production, but s/he can do that only when the relationship is strong and positive.

JRT is the least quantitative of the three J programs, and the best way to learn the concepts is through case studies. Each participant will bring in his or her own case study, which will cover a variety of situations. The TWI Service wanted to make sure that all relevant points of JRT were covered by cases studies so, in addition to the participants' case studies, they included four in the manual.

> Day 1: The Joe Smith Problem: Introduction of the Four-Step Method
> Day 2: The Tom Problem: The Importance of Getting the Facts
> Day 3: The Shipyard Problem: The Importance of Weighing and Deciding
> Day 4: The Woman Supervisor: Dealing with Change

Note the problem on Day 4 is titled "The Woman Supervisor Problem" and entails a change in policy that used women as supervisors as well as men. This definitely was a problem in 1944, but is no longer today. The situation that was to be examined is one of change. I have changed the problem to one where management decides to change their operation from traditional manufacturing to Lean Manufacturing. Instead of jobs being separated by types such as assembly, machining, welding, and so on, there would now be product lines where several different types of jobs might be on the same line. The changes here are significant and so there is much to talk about. In addition, it is a very common problem with which many people have to deal. All case studies are conducted according to a standard procedure, which helps the trainer and the participants stay on track (Figures 7.14).

Confidentiality

The case studies all involve personnel issues and some may be quite sensitive. One client did not want to have the participants use actual case studies because they believed that no such information should be discussed in a training setting. On the other hand, the issues will be discussed anyway and when done in a training setting, it is much more controlled. The disadvantage of purchased case studies is that they are never complete. If someone asks a question that is not covered in the written material, how is it answered? When a participant delivers a problem that s/he experienced, s/he will be able to answer any question about it since they lived through it.

JIT and JMT have forms that are used in the method, but there is only one form that may be used in JRT. That form should be collected at the door before everyone leaves because it contains personal information. The

Standard procedure

1. Ask supervisor to tell problem	Head of table. Does this involve you and somebody who comes under your direction? Have you taken action? Tell up to final action.
2. How problems come up	Where appropriate, stress you sensed or anticipated a change.
3. Get objective	Get from supervisor: Something to shoot at. May be changed. What do you want to have happen here? Does this problem affect the group? What net result do you want after you have taken action? Get group agreement.
4. Get facts	Supervisor first as he recalls them offhand. Review subheads with supervisor—USE CARDS. Get additional facts from group—USE CARDS.

5. Weigh and decide	Fitfacts—Look for gaps and contradictions with group
	
	Possible action:
	
	  What facts used?—From supervisor
	
	  Check practices and policies with supervisor
	
	  Check objective first with group, then with supervisor
	
	  Check probable effect on individual, group, production, with supervisor

6. Balance of case	Facts used (from supervisor)
7. Check step 3	Subheads—with supervisor Why? –How? –Timing?
8. Check step 4	Subheads—with supervisor When? –How often? –What?
9. Check objective	Supervisor
10. Foundations (if applicable)	Supervisor

(Thank supervisor and clear board except questions and steps)

Figure 7.14   JRT standard procedure.

form (Figure 7.15) does help people analyze the situation because they do not have to remember every detail and there is no need for them to keep the form at the conclusion of the session. All used forms should be collected and shredded.

Job relation training

Review sheet

1. How did this problem arise?

Sizing up/Tipped off/Coming to you/Running into you

2. What is the supervisor's objective?

3. What are the facts?

Review the record

Listen carefully to get the whole story.

Consider rules and customs

Talk with individuals

Get opinions and feelings

Facts:

Figure 7.15   JRT worksheet.

4. Weigh and decide

Look for gaps or contradictions

Do the facts fit together?   *Don't jump to conclusions.*

Consider how the facts affect each other

Check practices and policies

List possible actions

Consider the objective and the affect the action has on the individual, the group and production

Possible actions:

5. Take action

Are you going to handle this yourself?

Do you need help?   *Don't "pass the buck"*

Should you refer this to your supervisor?

Watch the timing of your action.

6. Check results

Look for changes in output, attitudes and relationships.

What effect did the action have on the person, the department, and production?

Did you accomplish your objective and improve production?

Figure 7.15 (Continued)   JRT worksheet.

7. What Foundations apply?

8. Could this problem have been prevented? How?

9. Did you accomplish your objective?

Figure 7.15 (Continued)   JRT worksheet.

Improved Relations: Beginning the Use of JRT

JRT is unlike JIT in that there are no specific jobs one can start working on directly after the training has been completed. It is also not like JMT, where someone will always have an idea for a better way of doing something. Once employees have learned how to use JRT, they simply have to start using it. Again, the foundations are what the focus should be and the four-step method used when required. All employees who have taken the 10-hour JRT program should be aware of how others who have taken it are using it, so they can monitor each other. This is especially true of the foundations. Of course, any personnel issues that do arise should be addressed with the four-step JR method and a person's supervisor can review how that was done. In order to keep the four-step method in people's minds until it becomes habitual, case studies (either real or fabricated) can be discussed at team or supervisor meetings. Everyone in the discussion must have been through the 10-hour JRT program for this to be effective. One HR manager said that whenever a supervisor approaches her about a personnel situation, the first thing she does is pick up the JRT pocket card. The supervisor knows by seeing that action that she will be asking for an objective, facts, and possible solutions.

Notes

1. Dinero, Donald A. 2011. *TWI Case Studies: Standard Work, Continuous Improvement, and Teamwork.* Boca Raton, Florida: CRC Press.
2. Reference JRT Foundations: Tell people in advance about changes that will affect them.

3. Refer to Appendix 8: Other Training Aids.
4. Training within Industry Service. 1943. *Job Methods Session Outline and Reference Material*. Washington, DC: War Manpower Commission Bureau of Training, p. 3.
5. Training within Industry Service. 1944. *Job Instruction Session Outline and Reference Material*. Washington, DC: War Manpower Commission Bureau of Training, p. 1.
6. Training within Industry Service. 1945. Training within industry report, 1940–1945. Washington, DC: War Manpower Commission Bureau of Training, p. 186.
7. Allan, Charles R. 1919. *The Instructor, The Man, and The Job*. Philadelphia: J.B. Lippencott, p. 361.
8. Training within Industry Service. 1944. *Job Instruction Session Outline and Reference Material*. Washington, DC: War Manpower Commission Bureau of Training, p. 3.
9. Two hours of training + three 1-hour coaching sessions = 5 hours for each of two sessions.
10. Note that any company should develop only one or two trainers to start with, since that will be sufficient to control the program as it starts. Developing six or eight trainers at the infancy of a program can cause it to grow too fast, resulting in confusion.
11. A missing piece is that the consultant acting as the Institute Conductor never returns to watch the trainer candidate deliver the 10-hour program to actual participants.
12. The most trainer candidates I ever had was four in one session, but even then there was not enough time to have everyone deliver the material to the point they were comfortable, practice JBSs, and comment and question in general.
13. I have modified the original JIT Trainer's Manual so that it follows the original material with the altered formats I am discussing here, and have also added helpful references.
14. One trainer who delivered the material particularly well said that he practiced by himself several times and then asked some participants to critique him. After that, he practiced again with the comments they had made and then delivered the material again to the group.
15. I avoid using the phrases "telling only" and "showing only" because most people combine showing and telling when they instruct. If I give a demonstration of "telling only," many people may think it is not relevant because it doesn't apply to them. In addition, they may believe the demonstration is not relevant and not say anything about it. I won't know who agrees with the motto and who doesn't.
16. Some people are ambivalent about saying that the trainer was at fault. Although it was obvious in this demonstration that the trainer delivered poor instruction, many people see themselves delivering similar instruction. This is the result you want because you will be showing them a method that overcomes these flaws and they will be more liable to accept it.

17. It should be noted that performing a job off-line is different than performing a job on a production line because they are two different environments. Several reasons a training area is different from a production area is that a production area may be noisier, may have more distractions, and will have more demands of getting and giving product. All those reasons may create stress and diminish proficiency.
18. Note that this may be the same person.
19. When a person is familiar with a job, waste can be difficult to see. JMT gives us a method to use to see waste that has been invisible to us.
20. Training within Industry Service. 1943. *Job Methods Session Outline and Reference Material.* Washington, DC: War Manpower Commission Bureau of Training, p. 6.
21. Training within Industry Service. 1944. *Job Relations 10-hour Sessions Outline and Reference Material.* Washington, DC: War Manpower Commission Bureau of Training, p. 18.
22. Ibid., p. 32.

Chapter 8

Sustaining the TWI Programs

> Once you climb to another level, you have to figure out how to sustain it.
>
> **Mary J. Blige**

The Ten Points

The biggest question I have had since I learned about the Training within Industry (TWI) programs is, "If the TWI programs are so good, why did people stop using them?" From my research and experience, I can see that TWI fell out of favor in the United States around the early 1970s. Although it has been used around the world, there are few places where it is actually thriving. For example, although it was used extensively in India, it is being reintroduced there now. This phenomenon is being repeated today, so one of the reasons for this book is to prevent it from losing favor again.

I have listed 10 points required to implement and sustain the TWI programs, and I have discussed each of them throughout this book and in other documents.[1] I include them here as a reminder (Box 8.1). All of these factors are required for both implementing and sustaining the programs. Absence of any one will not immediately sabotage a program, but the quality and use of it will degrade over time to the point where it will be discarded.

> **BOX 8.1   IMPLEMENTING AND SUSTAINING THE TWI PROGRAMS**
>
> 1. *Top management backing*: If TWI is not part of the overall strategy, it will fall into disuse. The TWI programs cost time and money and thus will be an addition to employees' jobs as they perform them now. The CEO and his/her staff should not only be aware of implementing the programs, but s/he should also make it a part of the overall strategy. If the CEO and staff believe it is just another training program, it won't have the backing it deserves. If, however, they realize that it is a problem-solving/culture-changing tool that forms a foundation for success, they will support it by making the correct decisions when questions arise about the programs.
> 2. *Management support*: Management must take responsibility for TWI. Since the programs take time, people must be given that time. Instruction and Methods Improvement must be considered to be a part of each employee's job. Budgets and schedules must be modified. The rewards will be great, but there is a price to pay. Most importantly, the results of the TWI programs should be a key performance indicator (KPI) for managers.
> 3. *Line organization participation*: Line management must participate since production is their responsibility. Everyone can benefit personally and professionally from using the TWI programs, but initially not everyone will be using them. However, when a group of employees starts using one of the programs, the line management (all supervisors) attached to that group should take an active interest by seeking metrics and results. After all, they are responsible for quality and productivity. They should also participate in a 10-hour session so they have experienced the programs. Unless a person has participated in a program, it is very difficult to have a deep understanding of it. This also changes it from being "the training department's program" to "our company's program." This is the way we do things now.
> 4. *Reporting of results*: If a gain is not recognized, it will fall into disuse.
> 5. *Appointment of a coordinator*: If everyone is in charge, no one is in charge. Someone must be assigned to coordinate the TWI effort. Typical tasks could include

a. Schedule and prepare for the 10-hour sessions
b. Schedule participants
c. Follow up with supervisors after the training to see how it is being used and to offer input as needed (includes helping set up an audit program)
d. Obtain and distribute results
e. Audit
f. Coach creation of JBSs
g. Coach instruction
h. Deliver the 10-hour programs if they have been trained to do so
i. Facilitate JBS groups (for getting consensus), MBS groups for getting consensus on larger JM improvements, and JR groups
j. Help people identify problems that can be solved or reduced through the use of the programs. The TWI coordinator would be a good candidate to learn and then train others in Program Development

6. *Quality institutes for participants*: There is no shortcut for learning how to use TWI. Discipline must be maintained in delivering the 10-hour programs so that quality does not diminish. As people become familiar with the programs, it becomes easier to attempt to "simplify" or "condense" them. A good trainer realizes that the programs are as simple and dense as they can be and any changes made when fitting them to an organization would be in augmenting them. In no case should any of the main principles be altered because that would decrease their effectiveness. For example, the amount of material a person can absorb in 10 hours over 5 days is greater in quality and quantity than can be absorbed in ten hours in one day. Periodic audits should be conducted on the 10-hour programs to verify that standardization is maintained. Non-standard training leads to non-standard work. If there is any question on the quality of the training institute (10-hour training), check with the Institute Conductor who trained the trainers.

7. *Complete coverage*: Optimal effectiveness requires everyone knowing and using the skill. The TWI programs can change an organization's culture but only if everyone knows about them and uses them. By "culture" we mean, "How we do things here." That requires that employees think in a certain way and use the same language.

If one person talks about *key points* and the other person does not know the term, communication will be difficult. The TWI coordinator should make sure that the entire organization receives the 10-hour programs and that refresher sessions are scheduled as needed.

8. *Coaching*: Feedback is required for learning and correct external feedback accelerates learning. Once an employee has been through a 10-hour program, she or he will require some coaching to enable proficiency in using the method. Having been through the 10-hour JIT program, for example, each participant will have been coached in creating and delivering a JBS. They will be knowledgeable but not proficient and thus coaching is required for the employee to sharpen his/her skills. Coaching is not 'telling' but rather helping the employee to strengthen weak areas of the skill. That means the coach must determine where the person is not strong and then decide how best to sharpen that aspect of the training.

9. *Correct use of the programs*: Use to make improvements and not as a "training filler." All training should be considered to be a problem-solving tool. We should not train for the sake of training but to address a particular problem. Therefore, when we measure the success of the training, we measure how well the particular problem has been solved and not the extent of the training. If the problem has been solved and we do not anticipate it returning, the training is no longer necessary and should be stopped.

10. *Conduct periodic audits*: Habits take time to form. Once everyone has been trained in a given job, it is usually necessary to follow up periodically and see that it is still being performed according to the standard. People who have done the job before another way will have habits that must be broken before new ones can be formed. Someone should be assigned to periodically view the job and determine that the standard is being followed. The training matrix is a good tool to use to keep track of this. If a person has performed the job according to standard several times over the period of a month or two, you can be fairly certain that the person will continue to perform it that way. One must not make the assumption that because someone has been properly trained in a job s/he will perform it according to the standard. If standards are important enough to spend time and money on training, they are important enough to follow up with an audit.

The Kirkpatrick Model

The Kirkpatrick Model[2] is a training evaluation model that can result in continued use of the TWI programs and thus continued successes. Much more can be learned about this model on their website, but I will include a brief overview here. The model describes how training contributes to results. People often question how they can justify their training programs. If you do not know how to justify training, you are not using training properly. Training should be used to achieve quantifiable results and once those results are achieved, the justification is obvious. The Kirkpatrick Model consists of four levels:

1. Reaction: The degree to which participants find the training favorable, engaging, and relevant to their jobs
2. Learning: The degree to which participants acquire the intended knowledge, skills, attitude, confidence and commitment based on their participation in the training
3. Behavior: The degree to which participants apply what they learned during training when they are back on the job
4. Results: The degree to which targeted program outcomes occur and contribute to the organization's highest-level results

Level 1 (Reaction) is measured by the feedback sheet completed on the last day of training and also by the discussion of "What I Learned." Two of the trainer's objectives are to have the participants see the need for the use of the program and thus want to use the program. If these objectives are met, then the participants will see the program in a favorable light.

Level 2 (Learning) is attained in the training itself since the TWI programs are based on a "learn by doing" concept. Every participant must perform the respective skill to show the trainer that s/he did in fact learn it.

Level 3 (Behavior) will be up to whomever is responsible for driving the program after the initial first week of training. That person could be the coordinator or the trainer. In training there is little worse than having a successful training session that is followed by not using what was learned.

In Job Instruction Training (JIT), before the week of training, a plan should have been made that would have identified an area for improvement. Some Job Breakdown Sheets (JBSs) may have been written during the week to be used in the demonstrations. Those should be finalized by getting consensus and the plan should be followed. Other JBSs that are necessary should be written, but be cautious to write only JBSs that are needed. The objective is to make an improvement and not to accumulate as many JBSs as possible. It is discouraging to spend time and effort writing a JBS that is not used. The only employees eligible to do the instruction are those who participated in the 10-hour JIT, but usually not all of them will be used as instructors. Anyone chosen as an instructor should start rehearsing the delivery of a JBS as soon as it has been finalized. Then the instruction should start and be followed sometime later with assessments. As more and more JBSs are written, more employees will be instructed and the program will become more familiar to everyone. If it is not possible to have all 20 participants involved initially, the participants who are not involved should be kept up-to-date on the activities.

Using Job Methods Training (JMT) does not have to be as formal as JIT because its use depends on how many ideas are initiated, and how often. There should be a formal continual improvement (CI) system of some sort to act as a clearinghouse for all the ideas and improvements, and management should be overseeing the entire effort. A caution here is to avoid setting goals for ideas. When JMT is started, it will probably fill a gap that has existed for a while. Employees have had ideas but they did not know how to handle their implementations. Once they are given a method, the result will be as if a dam burst. Once the floodwaters recede, however, ideas probably will not come as fast. Putting an artificial goal of "X" ideas per week or per month is counterproductive. If someone has several ideas in one time period, s/he may hold back on some in order not to miss a future time period. If someone does not have an idea for a given time period, s/he may make up something just to meet the commitment. The best way to encourage people to offer their ideas is to help them use JMT. Do not put any restrictions on it and accept all ideas for review and analysis. Sometimes the strangest ideas can result in great improvements.

Use of Job Relations Training (JRT) is the least structured of all the J programs because its use depends upon personnel issues arising. Even in the best run and most progressive organizations, personnel issues will occur. The supervisor, of course, will handle these by using the JRT method with input or feedback from his or her supervisor. There is no formal assessment to see that the method is being used properly since it depends on the two supervisors to do that. However, cases can be reviewed periodically in supervisors' meetings as reinforcement of the method and also to make everyone aware of important actions.

Level 4 (Results) is the level that is often missed. The trainer/coordinator and participants are very industrious in creating and using JBSs or MBSs, and the objectives that were initially stated are accomplished. However, results are not adequately measured, posted or published in any way. If results are not posted, results will be known only through word of mouth and that method is fairly unreliable. One part of the plan that was developed before the 10-hour session started was to determine how to publicize results. Publicizing results on a bulletin board or through a company newspaper gives credit to those who have made the improvements and also encourages other employees to get involved. In addition, it spreads ideas, which will spawn other ideas, as was discussed in Chapter 4.

Embedding into a Culture

This brings us to how success can lead to failure. When gains made from any TWI program are not published, the programs can still be used successfully. The people who use them learn how powerful the tools are and continue their use. Monthly, quarterly and yearly reports are written, but no mention is made of the link to TWI. These successes can continue for months or years. Sometimes a specific goal is in mind such as a yearly ISO or FDA audit. Other goals might be more ambitious such as the coveted Baldridge or Shingo Awards. Whatever it is, once it is achieved, there is celebration and then everyone goes back to work. What then leads to failure?

My clients all start using TWI successfully. My objective with any client is to enable them to use any of the J programs on a continuous basis. The intent is not just to train employees but also to embed the program into the organization's culture. I have a few clients who change the selected TWI program once they have achieved success with it. The changes include reducing the time from five days to three or less, using computer slides instead of a white board, eliminating some content such as the relationship to standard work and continual improvement, and changing other content, to name a few. It cannot be emphasized too much that changes to the TWI programs are welcome, but they must be vetted before they become permanent. Most companies that make such changes do not vet them and consequently, the programs either degrade or they are discontinued. CI is a good concept to follow, but it must be done with care. The TWI Service was always seeking improvements and they documented how many changes were made to each of the programs. However, whenever a change was suggested, it was thoroughly scrutinized and then thoroughly tested before it was released nationally. The Service had the advantage that they had access to a wide variety and number of places to "beta test" each change they thought was valid. Each suggestion was based on a reason, and the improvement had to satisfy that reason while not creating any problems before it was accepted.

The size of the company is not relevant because this happens to both single- and multiple-site facilities. These companies have existing training structures in place and JIT is melded in with that structure. For example, if they have a training department, they maintain the training department, but the trainers now use the JIT method. When the program is new to them, they follow the method scrupulously. Their plan of action is based on a quantified objective such as reducing training time, or scrap, or the number of faults in an ISO audit. Some clients begin to change the JIT format before they have achieved success throughout the entire organization, while others change the format after they have achieved success on a company-wide basis. I speak mainly about JIT because that is the first program with which most organizations start. Degrading or discontinuing JIT leaves little incentive to use JMT or JRT.

When I check on the clients who change JIT before they have achieved success on a company-wide basis (and this is not a scientific study), the reason is that management, especially senior management, was not fully involved. This is especially the case if senior management changes. The people who had the final say on strategy and finances were not fully aware of what JIT was accomplishing in the area of the facility where it was being used. Furthermore, those who did recognize the benefits were not able to convince senior management of the value of JIT. The objectives may not have been clearly and quantifiably stated, and the program itself was not part of the overall management strategy. The TWI Service recognized this phenomenon and they created a guide that its representatives would use to sell senior management. They learned that if senior management was not fully engaged, the programs might start out well, but they would deteriorate over time.

People who are not familiar with JIT often say that it takes too long to prepare to train someone (the JBS) and to conduct the training. What many people do not realize is that, overall, JIT takes less time than other types of training because JIT trains a person to do the job correctly the first time. Other types of training rely on the learner completing his training through trial and error, which results in scrap, rework, and/or injuries. Most of the time and cost associated with scrap, rework, and injuries should be allotted to the training budget; because if any of those three results are seen, the operator probably was not trained properly. People also see the JBS as just another document that only duplicates their work instructions. Often, operators do not follow the work instructions because they are incorrect and that is often due to their complexity. The JBS is a document that is vital to maintaining the quality of a JIT program in order to create the anticipated objectives. Senior management must be fully involved if TWI is to be used successfully on a company-wide basis. If only first-line supervision or middle management is involved, the TWI program(s) chosen will experience the boiling frog syndrome.[3] The quality of the program will slowly degrade and they will not notice because it will be so gradual.

The more puzzling case is that of the organizations that stop using a TWI program once they have achieved success with it throughout the entire organization. I have just anecdotal information, but some research has been done that explains why people may stop using methods that resulted in success. Francesca Gino and Gary Pisano wrote a paper titled *Why Leaders Don't Learn from Success*.[4] The central thesis of the paper is that people often analyze why they failed, but they do not analyze why they succeeded. An analysis of success is as important as an analysis of failure. Without a systematic analysis of why we succeeded, we can only make assumptions that our success was due solely to our actions. Often this is not the case, since external factors and luck play a part in whatever we do. Success often leads us to stop questioning and therefore to stop learning. This can result in over-confidence, which in turn leads to us believing we can improve what we have already done. If the changes are not based on the true causes of our success, but rather on what we think caused our success, the improvements will not happen. For example, from the hundreds of people who have attended my JIT sessions, a common remark is that on the fourth or fifth day the person now understands what I said on the first day. Five days is required for optimal absorption and retention. A new trainer may believe the success was due mainly to his ability to deliver the material and thus he can save some time by reducing the session to 3 days. A post-action analysis may also uncover details about which we were not aware, but which can help us with future endeavors.

Mandatory Actions for Success

In order to maintain the successful use of the TWI programs, there are some actions we should take. First we should follow the 10 points listed in Box 8.1. Those are required not only to implement a successful program but also to sustain it. Second, we should overcome our over-confidence. We may be the best, but if we are, and "rest on our laurels," competition will surely overtake us. Third, as we are using the TWI programs, we should focus on the cause and effect for the results. Many companies have several productivity improvement programs going on at the same time and sometimes it is difficult to identify the cause of a result. Being aware that this is necessary and doing it as soon as possible will make the task easier. For example, if excessive scrap is being produced during a machine setup, proper instruction can probably reduce the scrap rate. If nothing else changes, the benefit of proper instruction is obvious. However, if the setup is redesigned or enhanced at the same time, there must be some discussion on what benefit can be attributed to the redesign and what can be attributed to the improved instruction. It is much easier to do this initially than it would be several months later. The fourth action we should always take is to perform a post audit. This is a natural activity if we fail at something we have planned, but it rarely happens when we are successful. There are always many external factors that affect what we do. Some are within our control while others are not. The excessive scrap on the machine mentioned above has been reduced because we have instructed our operators well and we have redesigned the setup tooling. Could it be that the parts used have a tighter tolerance and that is the true reason for the reduced scrap? Just as we perform a root cause analysis for a problem, we should perform a root cause analysis for a success. Moreover, this post-action analysis should be systematic and incorporated into standard procedures. We want to continually improve, but those improvements should be based on evidence and reasons.

In order to use the TWI J programs, people must use basic skills such as proper communication (exchanging ideas and opinions), compromise and consensus building, analysis, questioning, collecting facts (not assumptions), teamwork, and listening, for example. The more we use a skill, the better we become at it and the more engrained it is in our mind. As a result, the TWI J programs can be seen as personnel development programs and for that reason alone, they should not be discontinued. The requirement is to have senior management understand and accept that.

Notes

1. Dinero, D.A. 2011. *TWI Case Studies: Standard Work, Continuous Improvement, and Teamwork.* Boca Raton, FL: CRC Press, p. xxi and Sustaining the TWI Programs, TWI Learning Partnership website. http://www.twilearningpartnership.com/SustainingTheTWIPrograms2.asp.
2. The Kirkpatrick Model. http://www.kirkpatrickpartners.com/OurPhilosophy/TheKirkpatrickModel.
3. The idea is that a frog will quickly jump out of a pot of boiling water, but will stay in a pot of cold water even after it has been heated to boiling. There is some controversy whether this is actually true and no one is willing to test it by killing a frog. The point is that when changes happen gradually, we often do not notice until it's too late. A major example is climate change. There are many sources that discuss this and I chose to reference it from Wikipedia. Boiling Frog. https://en.wikipedia.org/wiki/Boiling_frog.
4. Gino, F. and Pisano, G. 2011. Why leaders don't learn from success. *Harvard Business Review.* April. https://hbr.org/2011/04/why-leaders-dont-learn-from-success.

Chapter 9

The Future of the TWI Programs

> Without continual growth and progress, such words as improvement, achievement, and success have no meaning.
>
> **Benjamin Franklin**

A Look Back

The Training within Industry (TWI) programs were developed when the United States was in a crisis and, because they had limited time, money and personnel, the TWI Service placed requirements on how the overall objective of increased productivity was to be met. Each of the executives who comprised the nucleus of the Service knew what to do for an individual company, but productivity had to be increased for all defense contractors across the entire country. Therefore, whatever was to be done had to apply equally to any company and few assumptions could be made. Few companies had training departments and industrial engineering departments were not as common as they are today. Because time was a factor, whatever material was given to a company had to be straightforward enough that the average person could be quickly taught to disseminate it with successful results. At the close of the TWI Service in September 1945, a final report was written that described the five-year journey of the TWI Service. In the Preface, C.R. Dooley, the director of the TWI Service, wrote

> TWI work was undertaken to meet the specific objective of immediately increasing production for defense, then for war. No long range objective was set, but there are many implications for the development of the individual and for the improvement of the country's educational system.[1]

This is an important statement because it shows that although the programs were designed for wartime use, they knew that they would be as valuable in peacetime. Furthermore, although the stated objective was to increase productivity, another main objective was "development of the individual." Dooley goes on to quote Charles Mann, who acted as a consultant for the TWI Service and had been the chairman of the War Department's Committee on Education and Special Training during World War I:

> The training we give to the worker to do a good job *now* for war production can be more than an expedient means of getting the job done. It can be suitable to the individual and in line with his native talent and aspiration. Then it becomes education because the worker placed in the line of work he desires, and trained in accordance with his talent and aspiration, is a growing individual—mentally, morally, and spiritually, as well as technically. Training done from this point of view promotes production now and builds better citizens for a greater national stability afterwards.

These thoughts were ahead of their time and, because of the concept of the adjacent possible, the TWI programs remained a set of training programs that were thought to only increase productivity. Concepts such as a Learning Organization, Servant Leadership, team science, Human Performance Technology, Organizational Development, and meaningful purpose could not be shown to relate to TWI for two main reasons. First, most of these concepts had not been conceived during the time the TWI programs were being used widely. They were being developed around the same time the TWI programs were losing favor in the United States. Second, the TWI programs were thought mainly to be for process improvement and not employee development. Process improvement and employee development had been thought to be non-related until recently when some people began to recognize "the human side of Lean."[2]

Beyond the Original Intent

I have discussed some hypotheses about why companies stop using the TWI programs, but the programs not working as intended was never one of the reasons. Indeed, it has been shown that, when used properly, these programs work as intended in any company, in any country with any society. Since the programs were reintroduced into the United States at the beginning of this century, they have been shown to be as successful as they were in the 1940s. Even companies such as IBM, which praised them for their results during WWII, stopped using them, and then reintroduced them in 2005, find they are as successful now as they were then.[3] As people continue to use the TWI programs, they begin to see beyond the initial stated objectives and see relationships with the concepts mentioned above. The TWI programs are fundamental and when used as designed, they will produce the intended improvements in quality, productivity, safety, and cost. However, they also provide a necessary foundation for developing other necessary skills. Using the TWI programs develops skills that are necessary for higher-level programs and that, perhaps, is a better reason to continue using them. People see the structure of the TWI J programs and they see how the four-step methods can be used for instruction, process improvement and personal relations. Many people stop right there because, since their objective is to assist product designers, strategists, or CEOs, they believe TWI does not apply. They do not see how the TWI programs can help someone who is not making a physical product. They want a structured step-by-step program that they can give people to follow. That is akin to saying that there is a step-by-step program for love. There are many quantitative activities that product designers, strategists, and CEOs do that can be addressed with the J programs. I had a CEO in a JIT session who chose as his demonstration the writing of his monthly report to the board of directors. There are many jobs we think of as being 100% artistic or qualitative, but actually consist of much science and process. There is even much science to playing a musical instrument or a sport, which is why the BeLikeCoach organization uses the TWI programs.

Creativity

One issue that has been raised in the TWI discussion is that the important function of creativity is missing. How do we teach creativity? The answer is that we don't. The best we can do is prepare people to be creative and create an environment where they can become creative. John Adair's book *The Art of Creative Thinking*[4] gives several ideas on how to become a more creative thinker. Curiosity is one idea he states is important. We really can't teach people to be curious, but if people practice Job Instruction Training (JIT) and Job Methods Training (JMT), they will very definitely develop a questioning attitude, which is a requirement for creativity. JIT requires that we always have a reason for a Key Point and thus we continue to ask "why?" I can see this attitude develop in the five days I spend developing a coach. Initially people will accept any answer to "why" but once they understand that we must know the root cause, they continue to question on their own. Of course, JMT is much less subtle about developing this questioning attitude because we ask "why" for every detail we write. Once JIT and JMT have been learned well, the questioning attitude transfers to other activities, and people question more. They become more curious. Listening is another idea Adair cites and that of course is used in JIT and JMT, and emphasized in Job Relations Training (JRT). Another idea is to "sharpen your analytical skills," which is practiced in all three J programs, and "suspend judgment" is stressed in JRT, when we must get all the facts before making a decision. There are many other examples of how the basic skills developed in the J programs help us to be successful with higher-level activities.

Adair notes that creativity is "the power to connect the seemingly unconnected and this is enabled when we reach into a discipline different than our own."[5] A classic example of this is in the formation of the Just-In-Time concept of inventory control. Taiichi Ohno knew he wanted only enough inventory to build a product, thinking that excess inventory is waste, but he did not know how to implement the plan. Although he studied American auto manufacturers, all of them relied on significant quantities of inventory to maintain production, so the answer did not lie with them. After learning how American supermarkets restocked their merchandise, he used the same idea for Just-In-Time and thus he created "supermarkets" in the Toyota Production System.[6] He took an idea from an American retail food industry and used it in a Japanese industrial automotive industry, crossing over from one discipline to another.

The TWI programs were designed to teach all people how to transfer their knowledge, make improvements, and build and maintain strong, positive personal relationships. Does that mean that when a person has mastered the use of the three J programs s/he will blossom with creativity? What makes one person creative and another not? Why does one person span the adjacent possible while another does not? Why would Da Vinci design a helicopter even though he had no hope of being able to build a working model? Why do some people "connect the seemingly unconnected"? Johnson cites three main requirements for creativity.[7] The first requirement is intellect, but that cannot stand by itself because there are many intelligent people who are not creative. A second requirement is serendipity: being in the right place at the right time and being prepared. The final requirement is working at the edge of your discipline so you can easily cross over to other disciplines. Ohno was not so consumed with auto manufacture that he could not appreciate the operation of a retail grocery store. All three requirements work together in a symbiotic relationship. We must be smart enough to see relationships between ideas but we must also be prepared to accept those ideas when we see them. Da Vinci's interests and pursuits included invention, painting, sculpture, architecture, science, music, mathematics, engineering, literature, anatomy, geology, astronomy, botany, writing, history, and cartography.[8] At this point, I will add a fourth requirement for creativity: desire. A person can have a high intellect, be in the right place at the right time, and even be working across disciplines without being creative. This person must also want to see new concepts and want to make connections between the seemingly unconnected. There is no one course or program that will do this, but the TWI programs teach key skills that are basic requirements.

TWI as Foundational Skills

TWI was re-discovered when people, in the pursuit of Lean, were studying how the Toyota Production System works. It can be shown to be a foundational element of Lean. As I have continued to study TWI, however, I have seen that it is also a foundational element of several other disciplines. That is, when practicing TWI, one learns skills that are required in other disciplines. Furthermore, TWI helps people develop these skills often without them even knowing it. The following are some disciplines that are enhanced and supported by use of the TWI programs.

Health and Personal Wellbeing

I have already discussed how the TWI J programs align with the thesis of Self-Determination Theory (SDT)—Job Instruction enables competence, Job

Methods enables autonomy, Job Relations enables relatedness. When these three basic human needs are met, one's health and wellbeing are improved. A concept that is being discussed more frequently is that of meaningful purpose. Meaningful purpose for an individual can be defined as an understanding or belief of what the person exists to do. We work for pay, we work to be productive, and we work for a meaningful purpose. When we work for pay, we are working to survive and are at the bottom of Maslow's hierarchy (refer to p. 258), satisfying basic physiological and safety needs. When we work to be productive, we are in the mid-range of Maslow's hierarchy where we satisfy love, belonging and esteem needs. But when we work for a meaningful purpose, we start to achieve self-actualization. Most everyone has experienced interacting with someone who is working for pay. The job is done, but it may not be done well and certainly nothing is done beyond what is expected. People who work to be productive are more involved in their work and may even like their jobs, but when the day ends, so do any thoughts about the job. Those who have found meaningful purpose in what they do are almost one with the job. It is not "work" for them but it could almost be considered a hobby they get paid for. It would be difficult to have a scientific study to show that meaningful purpose affects wellbeing in a positive way, but there is much anecdotal information. Viktor Frankl wrote in *Man's Search for Meaning* a quotation paraphrased from Friedrich Nietzsche, "He who has a why can bear almost any how." Frankl was a psychologist who survived the Nazi concentration camps and saw evidence of this firsthand. Conditions were brutal, but the largest factor that affected someone's life was whether or not they thought they had something to live for. Not everyone knows his or her meaningful purpose, but even those who do, discovered it later in life. We must know our strengths, weaknesses, and character at a minimum before we can discover what our meaningful purpose is. Maslow's hierarchy might be considered WHAT is needed for maximum wellbeing and I propose that RAMP be considered HOW that is accomplished. RAMP[9] combines meaningful purpose with Self-Determination Theory:

R = Relatedness
A = Autonomy
M = Mastery
P = Purpose

The TWI programs assist us in developing relatedness, mastery and autonomy, which enable us to find our meaningful purpose.

The Learning Organization

Peter Senge popularized the concept of a Learning Organization in his book *The Fifth Discipline*.[10] A Learning Organization is one that facilitates the learning of its members and continuously transforms itself. Learning is an essential ingredient in making improvements and a Learning Organization has the capability of using the capacity of all of its members to improve products and processes. The TWI programs help develop factors that most people find difficult to create; yet they are necessary in order to create a successful Learning Organization. Senge states that a Learning Organization has five disciplines and those disciplines each have their own characteristics. Review Box 9.1 to see the characteristics of people who practice the five disciplines. Now review Box 9.2 to see the skills created when using the TWI J programs. Some characteristics such as gaining consensus are easy to align. Others, such as teamwork, are more difficult to see. Understanding others' jobs, getting consensus, exchanging ideas, and instructing someone all lead to better communication and teamwork. Teamwork is a characteristic that organizations find difficult to create, yet it is a natural by-product when the TWI programs are used correctly.

Coaching: BeLikeCoach

Another example is that of an organization named BeLikeCoach.[11] A research article written about the coaching (teaching) techniques of a UCLA basketball coach named John Wooden inspired the formation of this group. The BeLikeCoach founders saw problems with youth sports in their communities and as they investigated the problem, looking for solutions, they realized that the problems were common and were national in scope. Properly coached, youth sports teach many life lessons and build character and teamwork in children, and set a foundation for leadership. When the coaching is less than adequate, youth drop out of sports activities and lose this chance for development.

What makes BeLikeCoach's work interesting is that they analyzed the scholarly research on Coach Wooden to determine whether his coaching techniques were grounded in universal principles of teaching and learning. As Wooden's techniques were studied, they were found to be extremely close to those of TWI. In fact, the title of a book written by one of the researchers and one of Wooden's players is, *You Haven't Taught until They*

BOX 9.1 CHARACTERISTICS OF PEOPLE PRACTICING THE FIVE DISCIPLINES

Personal Mastery

- Continual learning
- Continually expanding ability to create the results they seek (= life-long generative learning)
- Acting on the desire to create
- Clarifying what is important
- Learning to see current reality
- Have a sense of purpose
- See current reality as an ally and not an enemy; know how to perceive and work with forces of change, rather than resisting those forces
- Inquisitive
- Committed, taking initiative
- Possess a broader and deeper sense of responsibility in their work
- Confident
- Commitment to the truth: Seeking to know why things are the way they are
- Able to separate what they truly want from what they have to do to get it; they don't concentrate on all the ways it cannot be done
- Able to integrate reason with intuition
- Able to recognize our connectedness to the world; i.e. how apparent external forces are actually interrelated to our own actions
- Increased compassion undermining attitudes of blame and guilt
- Commitment to the whole

Mental Models

- Be able to think through issues
- Continually questioning yourself
- Practice openness as an antidote to "games playing": Say (tactfully) what you are thinking
- Merit: Make decisions based on the best interests of the organization
- Aware of your own mental models: Recognize your views are based on assumptions: Reflection

- Being able to balance "advocacy" with "inquiry": Advocacy is saying what you believe, while inquiry is learning what the other person believes

Shared Vision

- Each person in the group is committed to having everyone else in the group having the same vision
- The vision reflects each person's own vision
- Teamwork is paramount because each person has the same vision
- Everyone is willing to experiment because they know the solution is possible even though they don't yet know what it is.
- People are willing to expose their way of thinking, give up deeply held views, and recognize personal and organizational shortcomings.
- People solve problems with the vision in mind
- Possess an openness and willingness to entertain a diversity of ideas.
- Are committed, not compliant (pages 204–5)
- Not only accept the vision, but also willingly (it's their own choice) want it
- People believe in a shared vision because they believe they can change the future

Team Learning

- Able to think insightfully about complex issues
- Remain conscious of other team members and can be counted on to act in ways that complement each other's actions
- Foster other teams by inculcating the practices and skills of team learning more broadly
- Master the practices of dialogue and discussion and can consciously move between them
- Know how to deal creatively with forces opposing productive dialogue and discussion
- Use the skills of reflection and inquiry
- Skilled in consensus building
- Conflict becomes productive
- Avoid or minimize defensive routines
- Possess a shared language

Systems Thinking

- Look into the underlying structures that shape individual actions and create the conditions where types of events become likely: Structure influences behavior
- Recognize that in order for you to succeed, others must also succeed
- Recognize how one's actions affect others; where and how one fits into the overall system
- Recognize that some actions create results gradually because of delays in feedback
- Avoid blaming others but attempt to find the real cause: Usually a structural explanation—We are all part of the system and the feedback process
- Look for a root cause where the solution prevents the problem from reoccurring
- Recognize the phenomenon of compensating feedback where working harder may only exacerbate the problem
- Recognize that short-term gains may result in long-term losses
- Recognize that small, well-focused actions may produce significant, enduring results, but usually these actions are not obvious
- Familiar solutions may not solve certain problems
- Recognize that the fastest solution or action is not necessarily the best
- Recognize that cause and effect may be separated in time and space
- Given time, one can realize multiple benefits from a single action and thus compromise might not be necessary
- Examine interactions with respect to the situation at hand and not with respect to artificial boundaries such as departments or functional areas

Have Learned.[12] This is a quotation often used by Wooden, which might mean he knew of TWI. Wooden always thought of himself as a teacher and his coaching techniques closely matched the JIT method. He referred to it as "whole-part-whole." He also had a penchant for continual improvement much like JMT and knew how to deal with people to build strong, positive personal relationships as is done in JRT.

BOX 9.2 J PROGRAM CHARACTERISTICS

Job Instruction

- Discuss jobs objectively
- Question details
- Question why actions are done
- Exchange views and ideas with other associates
- Organize work in a logical pattern
- Simplify jobs
- Instruct someone
- Understand jobs better
- Gain confidence
- Obtain a broader view of the operation
- Understand better jobs that others do
- Obtain consensus on specific jobs
- Question others' work methods
- Follow up

Job Methods

- Question details
- Question others' work methods
- Exchange ideas with other associates
- Implement ideas
- Compile data
- Present arguments
- Consensus on procedures
- Get a broader view of operations by considering all factors
- Collectively think of improvements
- Contemplate small focused improvements

Job Relations

- Look at facts, not assumptions
- Look at people's skills not on the job
- Treat people as individuals
- Consider entire situation, not just obvious incidents
- Addresses causes of problems and not symptoms
- Follow up
- Don't "pass the buck"
- Listen carefully to get the whole story
- Don't jump to conclusions
- Actions should help production

Once a strong relationship between TWI and Coach Wooden was established, the BeLikeCoach team took a strong interest in the Program Development module of TWI. This was particularly important because the goal of BeLikeCoach's work was to not have sport coaches mimic Coach Wooden but rather to create a replicable methodology for the design and sustainable development of educationally effective sport programs at any level of athletics, but particularly for youth sport programs. Further, in BeLikeCoach's mind, the sports environment seemed like 1940 with increasing societal concern about significant decline in youth sport participation. Just like TWI was used to build the defense industry, BeLikeCoach wanted to create a methodology based on learn-by-doing and train-the-trainer concepts that could stimulate problem-solving across all sports and levels of participation, especially youth sport.

The methodology created by BeLikeCoach is named Long-Term Program Development (LTPD) and it has been successfully tested with a variety of community- and school-based partners in sports across the United States. This successful testing is now leading to the creation of new career paths in physical education and sport management as athletic directors and coaches apply the LTPD skills to sports programs at the Olympic, collegiate, school, club, and recreational levels of sport.

Maslow's Hierarchy

It is interesting to note that Abraham Maslow published his classic paper "A Theory of Human Motivation" three years after the TWI Service began its work.[13] We should not be too surprised, however, since it has already been shown through Self-Determination Theory that the TWI J programs are closely linked to human motivation. The relationship between Maslow's Hierarchy and the TWI programs can be seen in Figure 9.1.[14] Merely by using these programs, basic needs, psychological needs and self-actualization needs are enabled.

The TWI programs offer any organization many benefits just based on its original design—improving quality, productivity, safety, and cost. But the greater benefits lie in those "by-products" that many people find so hard to attain.

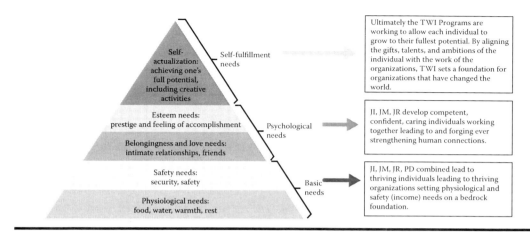

Figure 9.1 Maslows's hierarchy versus the TWI Programs.

Human Performance Technology

Human Performance Technology (HPT) is a field as broad as Lean and as a result I will only mention the connection here. Various people offer definitions of HPT, but one that can be used is

> Human Performance Technology is the study and ethical practice of improving productivity in organizations by designing and developing effective interventions that are results-orientated, comprehensive and systemic.[15]

Lean and HPT complement each other greatly and both have productivity improvement as the central goal. While Lean is based on process improvements, HPT is based on human interventions. As a result, TWI can be shown to be foundational for both disciplines. Lean focuses on the "hard skills," such as value stream mapping, flow, and standard work, but does recognize the "soft skills," such as employee development and motivation, while HPT focuses on the "soft skills," such as organization development, while recognizing the "hard skills," such as root cause analysis and Six Sigma. Much more information is available on the internet and in the HPT Handbook referenced in the endnote.

Risk Management

The practice of risk management is based on the adage "If you fail to plan, you are planning to fail."[16] Successful people recognize that it is important not only to operate effectively but also to anticipate as much as possible what can go wrong, and to take suitable actions for either prevention or control. The insurance industry is based on this, but risk management goes beyond what is commonly thought of as insurance. One risk management firm writes on its home page that "risk management involves following a deliberate set of actions to identify, quantify, manage, and monitor those events or actions that could lead to a loss."[17] In order for this to be done the people responsible for risk management must know what is actually being done in the organization. It is not enough to know what is being done, but it is required to know how it is being done in order to anticipate a loss. The only way to do that is for the organization to perform standard work at all levels of activity as described in Figure 9.2. As previously mentioned, the only way to achieve standard work is through standard training and the only standard training is JIT.

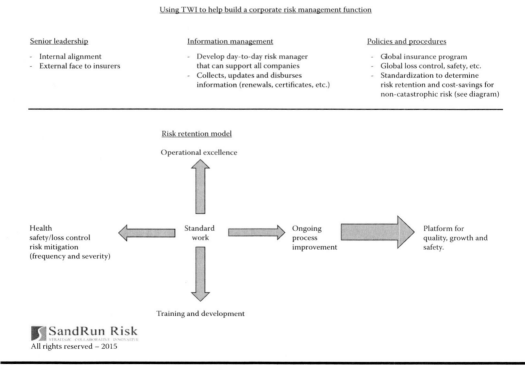

Figure 9.2 Risk management and TWI Programs.

Notes

1. Training within Industry Service. 1945. The "Training within Industry" report: 1940–1945. Washington D.C.: War Manpower Commission Bureau of Training, p. xi.
2. Dinero, D.A. 2013. Spanning the functional divide. *Performance Improvement*, Vol. 24, No. 9.
3. Dinero, D.A. 2011. *TWI Case Studies: Standard Work, Continuous Improvement, and Teamwork.* Boca Raton, FL: CRC Press, p. 41.
4. Adair, J. 2007. *The Art of Creative Thinking.* London: Kogan Page.
5. Ibid., p. 24.
6. https://en.wikipedia.org/wiki/Toyota_Production_System.
7. Johnson, S. 2014. *How We Got To Now: Six Innovations That Made the Modern World.* New York: Penguin, p. 252.
8. https://en.wikipedia.org/wiki/Leonardo_da_Vinci.
9. RAMP is an acronym coined by Lou Flaspohler, a physician and colleague who is studying effects on wellbeing.
10. Senge, P.M. 1990, 2006. *The Fifth Discipline: The Art and Practice of the Learning Organization.* New York: Doubleday.
11. http://www.belikecoach.com.
12. Nater, S. and Gallimore, R. 2006. *You Haven't Taught until They Have Learned: John Wooden's Teaching Principles and Practices.* Morgantown, WV: Fitness Information Technology.
13. Maslow, A.H. 1943. A theory of human motivation. *Psychological Review.*
14. Maslow's hierarchy of needs. https://en.wikipedia.org/wiki/Maslow%27s_hierarchy_of_needs Note that the article states that although Maslow never used the triangle to depict his theory, it is commonly used today. Figure 9.1 shows the triangle from this article enhanced by Lou Flaspohler.
15. Pershing, J.A. 2006. Human Performance Technology Fundamentals. In *Handbook of Human Performance Technology, Third Edition*, ed. Pershing, J.A. San Francisco: Pfeiffer, John Wiley & Sons, p. 6.
16. Attributed to Benjamin Franklin by http://www.goodreads.com
17. http://www.sandrunrisk.com

Appendix I: TWI Training Matrix

The following is the Training Matrix created and used by Jaime Portillo of Toyoda Gosei Fluid Systems, United States, located in Brighton, Michigan. Jaime is TGFS's Training Analyst and TWI JI Trainer.

264 ■ Appendix I

First Shift Team Four Team Training Matrix

"Failing to plan is planning to fail"

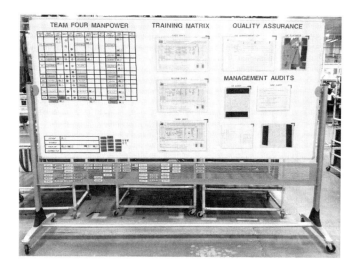

Appendix I

TG Fluid Systems
USA Corporation
Brighton, Michigan 48116 USA
Phone: 810.220.6161
www.toyodagosei.com

TG fluid systems team Training Matrix

The Team Training Matrix, also known as the Skills Matrix, is an up-to-date living document detailing the training needs of a particular group or team. The Training Matrix, as a visual management tool, outlines the process and employee capability of each team. Each matrix identifies who to train and on what procedure to help maximize employee flexibility and increase the competency of each process.

The Training Matrix is broken up into two groups; the Team Leader (TL) section, located along the top, detailing the competencies needed for each Team Leader, and the Assembly Technician section, describing the skill and capability of each operator. The TL Module levels, one through three, measures each Leader with a rating of a green "O" annotating the course or task was completed or a red "X" denoting the task has not been completed. The leader competencies were placed on the Skills Matrix to show a true representation of a "Team" Training Matrix and not solely an "operator" matrix. We quantify the operators within each team by rating them at a level zero through four, with the four qualifying them to train on the process.

Within the Assembly Technician section, Toyota Gosei Fluid Systems (TGFS) measures three core areas of training that provide us with the means to enhance the development of each individual as well as the organization. These areas are employee capability, process targets along with current percentages, and team manpower needs and requirements. To begin, all employee processes are broken down into a level of difficulty with `1` being the easiest process within the team and a rating of `4` symbolizing the more complex process within the team. All new hires begin on a level one process and continue rotating through level twos, threes and into fours as they become more proficient within their zones. The goal is for all team members to be 100% trained on all processes within their respective teams. The process targets and percentages annotated on the sample copy give us a baseline to work from and begin training.

The Team Training Matrix is placed on the Team Communication Board and training needs are communicated every day at shift startup by the Team Leaders and Trainers. This ensures that adequate preparation is done to ensure process stability as well as open communication to the Trainers and Trainees in letting them know of changes to their daily schedule in advance.

Appendix II: Cascading

We often need to instruct someone on a task that contains more information than the person can understand and remember in one training session. If we attempt to deliver information that requires several hours of instruction, it will be better received and retained if we deliver the material in smaller segments or "chunks." When we breakdown a job for a Job Breakdown Sheet (JBS), we may find that an Advancing Step is really a "freestanding" JBS. When we create the secondary JBS, we may find that an Advancing Step in that JBS actually has its own JBS and thus we must create a third-level JBS.

An example of this occurred when I asked a receiving clerk what demonstration he was going to use for the JIT session. He said that he wanted to do a fairly simple operation that people often have trouble with. The operation was receiving bulk oil from a tanker truck. I asked him what that entailed. He said his desk overlooks the staging area where the trucks park waiting to be serviced. When he sees a truck there, he walks out and welcomes the driver. He receives the Bill of Lading from the driver so he knows what the product is and has all the data for the order. He then reviews the receiving procedure with the driver and tells him what bay will be available for his truck. The clerk then goes back to his desk and calls an incoming inspector to tell him that an order has arrived. He tells the inspector what the product is because different products require different inspection materials. He then returns to the truck and guides the driver into the bay. He then attaches the pump to the truck's hose and unloads the oil. Once that's done, he removes the pump from the hose and completes the necessary documentation.

Now that I had a better idea of what he did, I watched him receive an order of oil, which took about 45 minutes. The result is the JBS Cascade 1 (Figure A2.1). This is obviously too large a job for a demonstration in the

No. 566

Job breakdown sheet

Operation: receive bulk oil
Parts: oil
Tools and materials: product receipt schedule; inspection receipt
Technical terms:
Common key points:

Advancing steps	Key points	Reasons
A logical segment of the operation when something happens to advance the work.	Anything in a step that might— 1. Make or break the job 2. Injure the worker 3. Make the work easier to do, i.e. "knack," "trick," special timing, bit of special information	Reasons for the key points
1. Welcome driver	1. Obtain bill of lading 2. Review procedure 3. Available bay	1. Confirm order data 2. Facilitates receipt 3. Don't have to return to truck
2. Contact Inspector	1. Product number	1. Inspector will bring the correct testing equipment
3. Guide truck		
4. Attach pump	1. Per JBS # 567	1. Most effective method
5. Unload oil	1. Complete receiving tag 2. Valve then motor	1. Receipt data visible 2. Prevents cavitating pump
6. Remove pump	1. Relieve pressure 2. Clean per JBS #568	1. Prevents oil splatter 2. Ready for next receipt
7. Complete documentation	1. Receipt form JBS# 569	1. Traceability
What	How	Why

Figure A2.1 JBS: receive bulk oil.

10-hour JIT session, but it was useful to create the JBS. Note that three Advancing Steps (#4, #6, #7) are involved enough to warrant their own JBS as shown in JBS Cascades #2, #3, and #4 (Figures A2.2 through A2.4). JBS Cascade #3 (Figure A.3) has an Advancing Step that requires an additional JBS Cascade #5 (Figure A.5). Attempting to instruct an individual

No. 756

Job breakdown sheet

Operation: disposal of hazardous material (solvents AK127; AC245; PQ952)
Parts: HazMat documentation sheet (HM769); solvent; HazMat container
Tools and materials: disposable cloths; safety glasses; rubber gloves
Technical terms:
Common key points:

Advancing steps	Key points	Reasons
A logical segment of the operation when something happens to advance the work.	Anything in a step that might— 1. Make or break the job 2. Injure the worker 3. Make the work easier to do, i.e. "knack," "trick," special timing, bit of special information	Reasons for the keypoints
1. Solvent into container	1. Verify container label 2. Wipe any spillage 3. Used cloths into HazMat container	1. Correct container 2. Fire prevention 3. Fire prevention
2. Complete HazMat doc sheet	1. All fields 2. Legible 3. Employee # and initials 4. File	1. Traceability 2. Others can read data 3. ID employee 4. Traceability
What	How	Why

Figure A2.2   JBS: disposal of hazardous material.

in all this information at one time would be problematic, but breaking it down into small JBSs reduces instruction time and increases retention and understanding.

The instruction can be performed in a variety of ways that best suits the learner and the environment, but here is one scenario. A person in

No. 569

Job breakdown sheet

Operation: documentation of bulk oil receipt
Parts: liquid inventory form (INV 248)
Tools and materials: bill of lading; PO; incoming inspection report (QA 109); black ink pen
Technical terms:
Common key points:

Advancing steps	**Key points**	**Reasons**
A logical segment of the operation when something happens to advance the work.	Anything in a step that might— 1. Make or break the job 2. Injure the worker 3. Make the work easier to do, i.e. "knack," "trick," special timing, bit of special information	Reasons for the key points
1. Complete Liquid Inv. Form	1. Verify BoL, PO, and QA 109 match 2. All fields	1. We received what we ordered per specs. 2. Describe traceability
2. Return BoL	1. Make copy 2. Sign initials and employee #	1. Original goes to driver 2. Traceability
3. Distribute Inv. Form	1. Attach copies of BoL, PO, QA 109 2. Purchasing 3. Acct. receivable 4. Receiving	1. Others can verify receipt 2. Close order 3. Pay supplier 4. Traceability
What	How	Why

Figure A2.3 JBS: documentation of bulk oil receipt.

the receiving department must know how to receive bulk oil. Because we want to start with the easiest task first, the first JBS we instruct could be JBS #756—Disposal of Hazardous Material Solvents (Figure A2.2). The next JBS might be JBS #569—Documentation of Bulk Oil Receipt (Figure A2.3). We start with the easiest tasks first to help the learner get familiar with the operation by association. He will watch us do the complete job as described in JBS 566 (Figure A2.1), but he will be learning only the simpler tasks. He will be watching what we are doing and so will be getting familiar with the other steps, but they will not be a formal part of his training at this time.

Advancing Steps #4 (Attach pump, Figure A2.4) and #6 (Remove pump, Figure A2.5) are similar, but AS#4 is a little less involved, so that may be the

No. 567

Job breakdown sheet

Operation: Purge and attach oil pump
Parts: oil
Tools and materials: spanner wrench
Technical terms:
Common key points:

Advancing steps	Key points	Reasons
A logical segment of the operation when something happens to advance the work.	Anything in a step that might— 1. Make or break the job 2. Injure the worker 3. Make the work easier to do, i.e. "knack," "trick," special timing, bit of special information	Reasons for the key points
1. Attach hose to purge pump	1. Both valves closed 2. Pump motor off	1. Control oil 2. Control oil
2. Purge pump	1. Open valve 1 2. Motor on 3. Open valve 2 4. 5 minutes	1. Oil flow to motor 2. Pump oil 3. Oil flow 4a. >5 min. wasted time 4b. <5 min. not sufficient for cleaning
3. Attach hose to oil pump	1. Both valves closed 2. Pump motor off	1. Prevents oil from entering pump 2. Doesn't start pumping oil
What	How	Why

Figure A2.4 JBS: purge and attach oil pump.

next JBS to instruct. Note that the tanker truck's hose is required for this. In order to instruct this JBS "off-line" one might obtain a length of similar hose to simulate that of the truck. Using that, you would not have to wait for the arrival of a truck to do the instruction.

Once these four JBSs have been learned, we can then progress to the final JBS #568. This could also be done with a simulation so the learner is very familiar with all the steps when an actual truck arrives. Depending on the learner, you may have him observe or you may observe him when the first oil order arrives.

No. 568

Job breakdown sheet

Operation: Clean oil pump
Parts: None
Tools and Materials: Solvent AK127; reservoir; disposable cloths; rubber gloves; safety glasses
Technical terms:
Common key points:

Advancing steps	Key points	Reasons
A logical segment of the operation when something happens to advance the work.	Anything in a step that might— 1. Make or break the job 2. Injure the worker 3. Make the work easier to do, i.e. "knack," "trick," special timing, bit of special information	Reasons for the key points
1. Attach reservoir	1. Both valves closed 2. Pump motor off 3. 1.5 gal. solvent	1. Control solvent 2. Control solvent 3a. >1.5 gal. overflow reservoir 3b. < 1.5 gal. insufficient cleaning
2. Clean pump	1. Open valve 1 2. Motor on 3. Open valve 2 4. 10 min.	1. Solvent flow to motor 2. Pump solvent 3. Solvent flows 4a. >10 min. wasted time 4b. <10 min. insufficient clean
3. Remove reservoir	1. Wipe down pump 2. Solvent into hazmat container per JBS 756 3. Wipe down reservoir 4. Cloths into HazMat container	1a. Easier to ID malfunctions 1b. Easier to handle 2. For traceability 3a. Easier to ID malfunctions 3b. Easier to handle 4. Fire prevention
What	How	Why

Figure A2.5   JBS: clean oil pump.

Appendix III: Self-Determination Theory

This is a theory describing human motivational behavior. That is, it attempts to describe why people do what they do. It states that everyone has three basic human needs that we are always, consciously or unconsciously, trying to meet. We tend to favor activities that satisfy these needs and avoid activities that hinder them. The three basic human needs[1] are

1. Competence: The need to manage yourself and your environment
2. Autonomy: Having the ability to choose your actions
3. Relatedness: To be understood, appreciated by, and connected to others

Competence is supported by Job Instruction Training (JIT) since when we truly know something, we have more confidence in doing it and thus we have more control over it.

Autonomy is supported by Jobs Method Training (JMT) since we now have the authority to change what we are doing to help ourselves.

Relatedness is supported by Jobs Relation Training (JRT) since we use the skills that lead to understanding and mutual respect.

For further information, go to http://www.selfdeterminationtheory.org.

Learning Organization

A Learning Organization is one that facilitates the learning of its members and continuously transforms itself.[2] In other words, it is an organization that is continually renewing itself through the learnings of its members. One of the characteristics is that all its members are encouraged to contribute to the effort. In order to be a Learning Organization its members must possess some characteristics and skills that are often difficult to acquire. There are five disciplines required of a Learning Organization, and Figure A3.1 lists the characteristics of the people who practice each one. Figure A3.2 lists characteristics and skills that are commonly acquired as one learns and uses the J programs. Comparing one list with the other will show the relationship between the TWI J programs and a Learning Organization, and how the former supports the latter.

Personal mastery
- Continual learning
- Continually expanding ability to create the results they seek (= lifelong generative learning)
- Acting on the desire to create
- Clarifying what is important
- Learning to see current reality
- Have a sense of purpose
- See current reality as an ally and not an enemy; know how to perceive and work with forces of change, rather than resisting those forces
- Inquisitive
- Committed, taking initiative
- Possess a broader and deeper sense of responsibility in their work
- Confident
- Commitment to the truth—seeking to know why things are the way they are
- Able to separate what they truly want from what they have to do to get it; they don't concentrate on all the ways it cannot be done
- Able to integrate reason with intuition
- Able to recognize our connectedness to the world; i.e. how apparent external forces are actually interrelated to our own actions
- Increased compassion undermining attitudes of blame and guilt
- Commitment to the whole

Mental models
- Be able to think through issues
- Continually questioning yourself
- Practice openness as an antidote to "games playing"—say (tactfully) what you are thinking
- Merit—Make decisions based on the best interests of the organization
- Aware of your own mental models—recognize your views are based on assumptions—reflection
- Being able to balance "advocacy" with "inquiry"—advocacy is saying what you believe, while inquiry is learning what the other person believes

Figure A3.1 Characteristics of people practicing the five disciplines in a learning organization.

Shared vision
- Each person in the group is committed to having everyone else in the group having the same vision
- The vision reflects each person's own vision
- Teamwork is paramount because each person has the same vision
- Everyone is willing to experiment because they know the solution is possible even though they don't yet know what it is
- People are willing to expose their way of thinking, give up deeply held views, and recognize personal and organizational shortcomings
- People solve problems with the vision in mind
- Posses an openness and willingness to entertain a diversity of ideas
- Are committed, not compliant (page 204-205)
- Not only accept the vision, but also willingly (it's theirown choice) want it
- People believe is a shared vision because they believe they can change the future

Team learning
- Able to think insightfully about complex issues
- Remains conscious of other team members and can be counted on to act in ways that complement each other's actions
- Fosters other teams by inculcating the practices and skills of team learning more broadly
- Masters the practices of dialogue and discussion and can consciously move between them
- Knows how to deal creatively with forces opposing productive dialogue and discussion
- Use the skills of reflection and inquiry
- Skilled in consensus building
- Conflict becomes productive
- Avoid or minimize defensive routines
- Possess a shared language

Systems thinking
- Look into the underlying structures which shape individual actions and create the conditions where types of events become likely—structure influences behavior
- Recognize that in order for you to succeed, others must also succeed
- Recognize how one's actions affect others; where and how one fits intothe overall system
- Recognize that some actions create results gradually because of delays in feedback
- Avoid blaming others but attempt to find the real cause—usually a structural explanation—we are all part of the system and the feedback process
- Look for a root cause where the solution prevents the problem from reoccurring
- Recognize the phenomenon of compensating feedback where working harder may only exacerbate the problem
- Recognize that short term gains may result in long-term losses
- Recognize that small, well-focused actions may produce significant, enduring results, but usually these actions are not obvious
- Familiar solution may not solve certain problems
- Recognize that the fastest solution or action is not necessarily the best
- Recognize that cause and effect may be separated in time and space

- Given time, one can realize multiple benefits from a single action and thus compromise might not be necessary
- Examine interactions with respect to the situation at hand and not with respect to artificial boundaries such as departments or functional areas

Figure A3.1 (Continued) Characteristics of people practicing the five disciplines in a learning organization.

Job instruction
- Discuss jobs objectively
- Question details
- Question why actions are done
- Exchange views and ideas with other associates
- Organize work in a logical pattern
- Simplify jobs
- Instruct someone
- Understand jobs better
- Gain confidence
- Obtain a broader view of the operation
- Understand better jobs that others do
- Obtain consensus on specific jobs
- Question others' work methods
- Follow up

Job methods
- Question details
- Question others' work methods
- Exchange ideas with other associates
- Implement ideas
- Compile data
- Present arguments
- Consensus on procedures
- Get a broader view of operations by considering all factors
- Collectively think of improvements
- Contemplate small focused improvements

Job relations
- Look at facts, not assumptions
- Look at peoples' skills not on the job
- Treat people as individuals
- Consider entire situation, not just obvious incidents
- Addresses causes of problems and not symptoms
- Follow up
- Don't "pass the buck"
- Listen carefully to get the whole story
- Don't jump to conclusions
- Actions should help production

Figure A3.2a "J" program characteristics.

Job instruction training

1. JIT—(5) 2-hour sessions
2. Write JBS$_s$
3. Get Consensus on JBS$_s$
4. Vet JBS$_s$
5. Rehearse JBS$_s$
6. Instruct with JBS$_s$—4-Steps
7. Audit operations, instruction, training
8. Expand area and repeat

Four steps
1. Prepare the worker
2. Present the operation
3. Try out performance
4. Follow up

Job methods training

1. JMT—(5) 2-hour sessions
2. Apply JMT 4-step method to local area
3. Document/publicize results
4. Repeat 2 and 3
5. Repeat 1–4
6. Repeat 5

Four steps
1. Break down the job
2. Question every detail
3. Develop the new method
4. Apply—vet, sell, implement

Job relations training

1. JMT—(5) 2-hour sessions
2. Apply JMT 4-step method to local area
3. Document/publicize results
4. Repeat 2 and 3
5. Repeat 1–4
6. Repeat 5

Four steps
　Determine an objective
1. Get the facts—include opinions and feelings
2. Weigh and decide—consider all possible actions
3. Take action—need help or authority? Timing
4. Check results—how soon; how often; changes in output, attitude, relations
Did you accomplish your objective?

Foundations:
- Let each person know how they are getting along
- Give credit where credit is due
- Tell people in advance about changes that will affect them
- Make the best use of each person's ability
- Treat all people as individuals
- Use empathetic listening
- Have high expectations, but accept and deal with failure
- Don't jump to conclusions
- Don't "pass the buck"
- Monitor actions with respect to production

Figure A3.2b Using the "J" Programs.

Notes

1. http://www.huffingtonpost.com/david-sze/researchers-determine-the-three-ways-to-well being_b_7512510.html.
2. http://en.wikipedia.org/wiki/Learning_organization.

Appendix IV: Script for Using JIT for Instructing How to Tie the Fire Underwriter's Knot

This script does not have to be followed exactly, but there should be no significant variations. That is, the content will vary with the Learner, but the concepts should be the same. Phrases in **bold** are critical to reproduce.

Instructor	Good morning. My name is Jack Brown. How are you doing today?
Learner	Fine thank you. My name is Jill Carpenter.
Jack offers his hand to shake hands.	
Instructor	Welcome to the electrical sub-assembly department. Did you have a good weekend?
Learner	Yes. It was very nice. Thank you.
Instructor	What did you do?
Learner	I have some relatives visiting from out of town and we all went to the park for a picnic. It was fun because I haven't seen them in such a long time.
Instructor	Picnics are a lot of fun—especially with family—and family that you haven't seen in a while. I hope you are all refreshed and ready to go. Where did you work last?
Learner	I was a material handler in shipping, but I've got an electrician diploma, so when I saw this opening I thought it might be a better fit for me.

Instructor	Well, we do all the electrical sub-assemblies for the rest of the company, so you'll be able to see all the products we make. You can see that this is a small department. We have only ten technicians so everybody has to be able to do all jobs.
Learner	That's great.
Instructor	As the supervisor in the department it is my responsibility to train you for all the jobs. Now we're going to start with a job that is common to many of our products. It's called the Fire Underwriter's Knot. What do you know about tying knots?
Learner	Not very much. I can wrap a package, but I don't know many different knots.
Instructor	That's fine. If you can tie a package, you can tie this knot. Here's where it's used. I know it may not look important but it acts as a strain relief in all of our fixtures. When someone pulls on the wire, the force goes to the knot and not the terminals. If the knot weren't there, the terminals might loosen, the wires could come loose and create a short. That might cause sparks and a fire.

The Instructor holds the fixture and then disassembles it to show the knot.

Instructor	Now I'm going to show you how to tie the knot. I'll go over it a few times and then have you try it. OK?
Learner	OK.
Instructor	Stand right here so you can see over my shoulder. Can you see OK?

The Instructor sits at the workbench and sees that the Learner is looking over his shoulder.

Learner	Yes, I can see fine.
Instructor	**This job has five Advancing Steps (AS). The first AS is to untwist.**

The Instructor lays the wire on his hand to approximate 6" and then untwists the wire about 6".

Instructor	**The second AS is to make a right loop.**

The Instructor makes a right loop.

Instructor	**The third AS is to make a left loop.**

The Instructor makes a left loop.

Instructor	**The fourth AS is to pass the right end through the right loop.**

The Instructor passes the end of the wire that is on the right directly through the right loop.

Instructor	**The fifth AS is to tighten.**
The Instructor takes both ends of the wires making sure they are even, and he pulls the wire to tighten the knot. In doing so he pulls the knot down slightly so that the wires are extended.	
Instructor	**So, the ASs are untwist, right loop, left loop, right end through right loop, tighten. Now I'll do it again and tell you the Key Points.**
The Instructor lays down the knotted wire and picks up an unknotted piece.	
Learner	OK
Instructor	**This job has five ASs. It has one Common Key Point, which must be done with each AS. That Common Key Point is to hold the wire at the end.**
The Instructor holds one wire at the end with his thumb and index finger and shows it to the Learner.	
Instructor	**The first AS is to untwist. That step has one Key Point (KP), which is 6".**
The Instructor lays the twisted wire on his palm, grabs the wire about 6" from the end and untwists the wire.	
Instructor	**The second AS is to make a right loop. That step has one KP, which is to pass the wire in front.**
The Instructor makes a right loop and shows the wire going in front of the vertical strand.	
Instructor	**The third AS is to make a left loop. That step has three KPs. The first KP is to pull the wire forward. The second KP is to pass the wire underneath, and the third KP is to pass the wire behind.** So the three KPs are forward, beneath, behind
The Instructor makes the left loop slowly and then undoes the loop. He then slowly does all three KPs again, individually, making sure the Learner can see. The sequence would be to say the KP, then do it, then say the second KP, then do it, and so on.	
Instructor	**The fourth AS is to pass the right end through the right loop. This AS does NOT have a KP. The fifth AS is to tighten. This AS has three KPs. The first KP is to hold the ends even. The second KP is to position the knot, and the third KP is to tug on the knot.** So the three KPs are ends even, position, tug.
The Instructor slowly performs each KP making sure the Learner can see what he's doing. The sequence would be to say the KP, then do it, then say the second KP, then do it, and so on.	
Instructor	Do you have any questions?

Learner	No, I think I've got it.
Instructor	OK. I'm going to do it one more time and this time I'm going to tell you the reasons for those KPs. There are five ASs. There is one common KP. That KP is to hold the wire at the end. I want you to hold the wire at the end because it makes it a lot easier to manipulate the wire and make the loops. The first AS is to untwist and it has one KP, which is 6″. I want you to untwist the wire 6″ because that will give us the proper length leads for subsequent jobs. If you untwist much less than 6″ there will not be enough wire for later operations. If you untwist more than 6″, it won't look good and we'll have to re-twist some. In either case, we'll have rework.
The Instructor performs Step 1.	
Instructor	The second AS is to make a right loop and that has one KP, which is to pass the wire in front. The reason for that is that if we pass the wire behind, we'll have a knot, but it won't be the right one and we'll have to start over, which is a time waste.
The Instructor performs Step 2.	
Instructor	The third AS is to make a left loop and this step has three KPs. The first KP is to pull the wire forward. I want you to pull the wire forward because that makes it easier to make the loop. You would do this after you tied these knots for a while, but I wanted to show you a trick to make it easier from the start. The second KP is to pass the wire underneath. I want you to pass the wire underneath because that will give you the correct knot. Again, you'll make a knot, but it won't be the correct one. The third KP is to pass the wire behind. There are two reasons for this. The first is to make the correct knot and the second reason is to put the wire in the correct position for the next step. Are you OK so far?
Learner	Yes, I think so.
The Instructor has performed each AS and each KP slowly as he says them	
Instructor	The fourth AS is to pass the right end through the right loop and that step does not have a KP.
The Instructor performs Step 4.	
Instructor	The fifth AS is to tighten the knot. This step has three KPs. The first KP is to hold the ends even. It's important to hold the end even because all of our fixtures are symmetrical and require both wires to be the same length. If one length is longer than the other, it would have to be trimmed, and we want to avoid rework. The second KP is to position the knot. We position the knot because we want to make

	sure the leads are long enough to go into our fixtures. It's very easy to tie the knot at the end of the wires and then the leads would not be long enough to attach to the terminals. The third KP is to tug on the knot. We tug on the knot because we're using cloth-covered wire and it has a tendency to "relax" and loosen. Therefore, we must "set" the knot so it doesn't unloosen and we do that by tugging on it. If you didn't "set" the knot, at least half of them would have to be reworked.

As above, the Instructor slowly performs the KPs as he is talking, making sure that the words coincide with the actions.

Instructor	Do you have any questions?
Learner	No, I think I've got it.
Instructor	Do you want to try it or do you want me to tie another one?
Learner	I want to try it.
Instructor	OK. Here's a wire.

The Instructor hands the Learner a wire. Notice that he did NOT ask her to say anything since for now he just wants to see if she can tie the knot correctly. The Learner begins to tie the knot. Most will talk as they tie the knot; some will not. As they tie the knot, the Instructor will correct anything the Learner does or says as soon as it happens. The following is what a typical Learner would say during the first interaction.

Learner	Well, I know I've got to measure 6" and then I straighten the wires and make a loop. I've got to bring the wire forward and pass it through the loop. Now I've got to get the ends even and tug on it.

The Learner slowly ties the knot and can be seen thinking what she has to do. When she's done, she hands the wire to the Instructor.

Instructor	That's a good job. You know how to tie the Fire Underwriter's Knot. I'd like you to do it once more and this time tell me the steps.
Learner	I don't think I know what the steps are.
Instructor	Well, I know you can tie the knot, so do it again and this time we'll just label what you do. How many steps are there?
Learner	There are five steps.
Instructor	OK. Start tying the knot.

The Learner starts to tie the knot by measuring 6" on her hand and then untwisting the wire.

Instructor	OK. What did you do?

Learner	I measured 6" and untied the wire.
Instructor	OK. The first AS is to untwist. Six inches is a KP. Continue.
The Learner makes the right loop.	
Learner	Now I make a right loop.
Instructor	That's correct. That's the second AS.
The Learner slowly makes the left loop.	
Learner	Now I pull the wire forward and put it behind.
Instructor	That's right, but what did you do?
Learner	I made the left loop.
Instructor	That's right. That's the third AS—make a left loop.
The Learner performs the fourth AS.	
Instructor	What did you do there?
Learner	I put the right end through the right loop and that's the fourth step. Now I've got to tighten the knot. I hold the ends even and tug on it.
The Learner performs all three KPs as she does step 5.	
Instructor	You just did the fifth AS. What is it?
Learner	To tighten the knot.
Instructor	That's correct. So you not only know how to tie the knot, but you also know the five ASs. I would like you to tie another knot and this time tell me the ASs and the KPs. How many steps are there?
Learner	There are five ASs. The first is to untwist the wire and the KP is to untwist it 6".
As the Learner talks she performs step 1.	
Instructor	Did I ask you to hold the wire in a specific way?
Learner	Oh, yes. I should hold it at the end because that makes it easier to manipulate.
Instructor	That's correct. Continue.
The Learner makes the second AS—a right loop.	
Learner	Now I've got to make a right loop and make sure the wire is in front.
Instructor	That's right. Making the right loop is the second AS and the KP is to put the wire in front.

The Learner pulls the wire forward and makes the left loop.	
Learner	The third AS is to make a left loop and I've got to pull the wire forward.
Instructor	That's correct. How many KPs are there?
Learner	There are three KPs. I've got to pull the wire forward and pass it behind.
Instructor	What's the third KP?
Learner	Oh, it's to pass the wire underneath.
The Learner performs AS #4.	
Learner	Now I pass the right end through the right loop. And now I tighten the knot.
The Learner performs AS #5 by doing all three KPs.	
Instructor	OK. What's the fourth AS and what are the KPs?
Learner	The fourth step doesn't have any KPs.
Instructor	That's correct! What's the fifth AS and what are the KPs?
Learner	The fifth step is to tighten the knot and the KPs are to hold the ends even and tug on the knot.
Instructor	That's correct, but how many KPs does the 5th AS have?
Learner	Yes, that's right. The fifth AS has three KPs and the last one is to make sure the knot is in the correct position.
Instructor	That's great, Jill! I would like you to tie one more knot and this time tell me the ASs, the KPs, and the Reasons for those KPs.
The Learner performs each step and then tells the Instructor what she's done.	
Learner	OK. The first AS is to untwist the wire, but there's a common KP to hold the wire at the end.
Instructor	That's right. Why?
Learner	It makes it easier to make the loops. The KP for the first AS is to untwist 6" so we get the right length of wire for the next operation.
Instructor	What happens if there's too much or too little?
Learner	If there's too much we'll have to rewind it and if there's not enough we'll have to re-tie it. In either case, it's rework.
Instructor	That's right. Continue.

Learner	The second AS is to make a right loop and the KP is to pass the wire in front. That gives us the correct knot. The third AS is to make a left loop and the KPs are to pull the wire forward and pass it behind. Passing it forward makes it easier to go behind and going behind gives us the correct knot and sets us up for the next step.
Instructor	That's correct. Do you remember how many KPs there are in AS #3?
Learner	There are three.
Instructor	Do you remember what the third one is?
Learner	No I don't.
Instructor	Well, you did it, but it's important to know what you did and why. Remember "forward, beneath, behind?"
Learner	That's right. I have to go under the wire so I get the correct knot.
Instructor	That's correct. Continue.
Learner	The fifth AS is to tighten the knot and that has three KPs. The first is to hold the ends even because the fixture terminals are symmetrical. If the leads aren't even, we'll have to trim one and that's rework. The second KP is to position the knot so we have enough length of the leads. The third KP is to tug on the knot to set it so it doesn't loosen before the next operation.
Instructor	That's very good, Jill. Do you have any questions?
Learner	No, I think I can do it.
Instructor	Are you ready to tie some by yourself?
Learner	Yes. I think I'm ready.
Instructor	Great. I'll leave you alone for a while and you can continue to tie the knot. I'll be back in about 5–10 minutes to check on you and see how you're doing. If everything is still OK, I'll let you go for an hour or so. If everything is still ok, I'll show you how to wire a fixture. In the meantime, if you have a question and I'm not around, ask Jack Hill. I'll introduce you now.
	INSTRUCTION HAS CONCLUDED

Appendix V: Creating and Using a Job Breakdown Sheet

Creating a JBS

Create a Draft

1. Watch an **expert**: To get as many Key Points as possible.
2. Watch at the **job site**: So you can recordxactly what is done, not what someone thinks is done.
3. Get **Advancing Steps first**, then find Key Points and Reasons—to give you a framework and so you don't get slowed down debating Advancing Steps versus Key Points.
4. If you have never seen the job before, ask the expert to **do the job once** so you can see what it entails.
5. Have the expert **perform the job several times** completely because you might see some action during one repetition that you missed during another.
6. When writing down a JBS, use **an erasable writing instrument** because you probably won't write exactly what you want the first time. When possible, have another person do the writing. This lets you concentrate more completely on watching the job.
7. **Process** for creating the First Draft:
   Advancing Steps:
   a. Having seen the job, mentally break it down into Advancing Steps—rough cut
   b. Ask the expert to perform the job and to stop when you ask him/her to stop—that will be when you believe s/he has done an Advancing Step

c. Ask the expert—"What have you done?"
"Did it advance the work?"
"Is it an Advancing Step?"

Use the answers and ask additional questions as required to determine whether the action is an Advancing Step. Use other criteria as required to determine if the action is an Advancing Step—is it WHAT is being done? Would the job stop if this action were not done? Could the job be safely and correctly handed off to another person after this action? Does it advance the work or does it *only* pertain to safety, productivity, quality, or cost?

d. Continue getting the remaining Advancing Steps, action by action. Review all the Advancing Steps as you have written them. Do they seem reasonable; do they make sense?
e. If Key Points are obvious and you think they might be forgotten, write them down at this time. Don't get "sidetracked" by getting all the Key Points.

Key Points

f. Have the expert perform the job again to document Key Points.
g. Ask the expert to perform each Advancing Step, one at a time, and then stop.
h. After each Advancing Step has been performed, ask, "Did you do anything that would make or break the job, injure yourself, or make the job easier? That is, is there anything that you did that affects quality, safety or productivity?
i. Discuss the action that the expert mentions and also verify that it is a Key Point and not an Advancing Step.
j. Keep asking the question until the expert says, No."
k. If you see something that you believe is a Key Point, but that the expert" did not mention, ask, "What would happen if you …?" "Why did you …?" Would it be easier if you …?" Often an expert may perform a Key Point subconsciously.
l. Once you have the Key Points for an Advancing Step, write down the **Reasons.**

Get Consensus from Other Experts

An "expert" is someone who is proficient in performing the job. Since you created the draft by watching only one expert, you should now get input from other experts in how they do this job. It's easier to comment on a JBS than it is to create one, so this method makes it easier for the other experts. That is, each expert does not have to write a separate JBS. You will also get improved input if you meet with all the experts at one time. It is easier to exchange ideas when people are speaking face to face. The final product will be a JBS that all agree on. Note that this "consensus meeting" should be limited to 6–8 people since more than that may result in a nonproductive meeting. If there are many experts, say 20 or so, you could have two meetings and then compare the JBSs from each one.

Verify the JBS by Training a Novice

Experts may take for granted some actions that people unfamiliar with the job may not know. Before you start using the JBS for general training, train someone who knows nothing about the job to see if anything has been omitted.

Practice the Delivery

Delivering the material from the JBS smoothly and correctly is advancing for successful training. Practice delivering the training until you are confident in its delivery.

Using a JBS

Notes on Delivering Job Instruction

Have both the Job Instruction reminder card and the Job Breakdown Sheet readily available for reference and then follow the Four Steps:

Step 1: Prepare the Worker

- Put the person at ease.
- State the job and find out what the person already knows about it.
- Get the person interested in learning about the job.
- Place the person in a correct position.

Step 2: Present the Operation

- Tell, show, and illustrate one ADVANCING step at a time.

    IDENTIFY & NUMBER each ADVANCING Step & Key Point

    Say, "There are **X Advancing Steps** in this job. The **first Advancing Step** is …" [Do not say anything else while performing the first step, but note that you must also perform all key points when doing the ADVANCING Step.]

    Then say, "The **second Advancing Step** is …" [Do not say anything else while performing the second step, but also include any key points.]

    Continue with this approach until the job has been completed.

- Stress each Key Point

    Perform the job again stating the Advancing Steps and the Key Points. Say, "There are **X Advancing Steps** in this job. The **first Advancing Step** is … This **Advancing Step** has **Y Key Points**. The **first Key Point** is …" [Be brief but complete.] Say, "The **second Key Point** is …" [Be brief but complete.] Continue until you have told the trainee all of the Key Points for Step 1.

    Then say, "The **second Advancing Step** is … This Advancing Step has **Z Key Points**. The **first Key Point** is …" [Be brief but complete.] Continue with this approach until the step and then the job have been completed.

    If an Advancing Step does not have a Key Point, tell the trainee that there is no Key Point for that step.

- Explain reasons

    Perform the job a third time as you did for Key Points. Explain the reason for each Key Point when you state the Key Point. Here we are more concerned about understanding than brevity. Make sure the learner understands before moving on to the next Key Point.

- Instruct clearly, completely, and patiently, but do not give more information than the person can master.

    Use the Job Breakdown Sheet as a guide to limit your conversation. For Advancing Steps, say only what is written.

    Use your judgment based on both the job and the trainee when deciding how many times to repeat the job completely. It is best to perform the job three times as stated above because the learner has more opportunities to see the entire job. As simple as the job appears to the instructor, much information may be new to the learner. If done only twice, the reasons would be given with the Key Points. If done more than three times, the Advancing Steps and the Key Points should be enumerated and the reasons given each time.

Do not give more information (verbal and visual) than the person can absorb.

Do not give irrelevant information.

Continue until you believe the person can do the job, and then ask the learner if he or she would like to try it. If they decline, repeat the job again, with ADVANCING Steps, Key Points and reasons as was done during the third iteration.

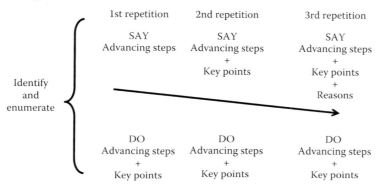

Step 3: Try Out Performance

- Have the person do the job—correct errors.
    - Correct errors *immediately*, so the person does not have a chance to learn an incorrect method.
- Have the person repeat the job and explain each Advancing Step, Key Point and reason. Make sure the person understands.
- It is usually best to have the person perform the job once without saying anything because that is easiest.
- People often confuse Advancing Steps with Key Points; so make sure the person enumerates them separately. Therefore, it is usually a good idea to have the person perform the job three times at least.
- Continue until you know the person knows.

Appendix V

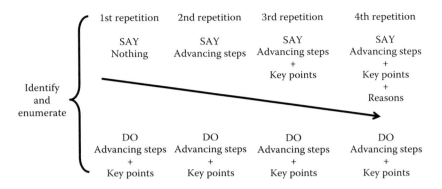

Step 4: Follow-Up

- Put the person on his or her own. Designate to whom they can go for help.
- Check frequently. Encourage questions.
- Taper off extra coaching and close follow-up.

> If the person hasn't learned, the instructor hasn't taught.

Appendix VI: The Philosophy behind TWI

We must know how they thought if we really want to do what they did.

Objective Is to Make Product

What we're all trying to do—everybody—is make some product to create a livelihood. That product can be anything we make no matter if we're a dentist, a lawyer, a welder, an assembler on a line, a CEO, and so on. Usually, we're in competition with someone else, so if our product is made resulting in the best quality, productivity, cost and safety, we can more easily sell it to others.

Develop Individuals

The prime objective must be met, but if individuals are not developed mentally, morally, and spiritually, as well as technically, the nation's stability cannot be assured. People want to be productive and thus most (80%) of problems are caused by lack of or poor training.

Identify and Solve Problems

Define a Problem

A problem is anything that interferes with our productivity. What we have to do is identify anything that interferes with us making that product and correct it. A problem can be physical, process-based or based on personal relationships.

Thus, a problem is anything that interferes with *productivity, safety, cost*, or *quality*.

The TWI service recognized that we can always improve, so we have to solve the problems that we recognize and also those that we do not yet see.

Problems That We Recognize

Program Development—a general problem-solving program based on the Scientific Method. They called it the Engineering Method, and often people refer to it as PDCA, but it contains any number of steps:

- Identify a problem
- Hypothesize a solution
- Try the solution
- Measure the results
- Modify as appropriate

All Problems Are Caused by People or Machines/Systems

If people: Use JIT or JRT
If machines/systems: Use JMT

Problems That We Cannot See

Other problems are embedded so deeply in our habitual processes we can't see them. These problems require analysis, which begins with relentless questioning.

Start with JIT

Find the expert who knows how to do the job best and break it down. When we do this we must get consensus so that our method is the best we collectively know at this time. That, by itself, will identify and solve some hidden problems.

Teach Everyone JMT

Everyone has ideas on how to change, which they think are improvements. They must know how to vet their idea to determine that their idea is, in fact, an improvement. Everyone must know how to vet, sell, and implement his idea.

Teach Everyone JRT

Many people cannot solve "people problems" but they can be taught to use JRT. The real strength of JRT lies in preventing problems from occurring in the first place. This is done best by the opposing sides. A moderator should be a last resort. In addition, JRT outlines the Scientific Method very clearly, so it can be used for any other type of problem.

Appendix VII: Training within Industry

Job Instruction Training

Planning Questionnaire

1	What improvements are you intending JIT to achieve? Include: Area or department included, e.g. reduce scrap from Product A or Department B Current status quantified, e.g. scrap 17% Desired future state, e.g. scrap <5%
2	What other improvements would you like achieved?
3	To what extent is management involved? If not fully involved now, when?
4	How do you plan to spread the use of JIT?
5	Who is to participate in the 10-hour sessions and in what order?
6	Is a trainer needed to deliver the 10-hour program? Who will be a trainer?
7	Who will be coaches?
8	Who will be the JIT (or TWI) coordinator?
9	What will be his or her duties?
10	How will JBSs be created and used?
11	Who will write JBSs?
12	How will JBSs be approved? By whom?

13	How will JBSs be documented and changed?
14	How will the Training Matrix be used? By whom? Posted or in file only?
15	How will results be reported? By whom?
16	What audits will be conducted? By whom?

Appendix VIII: Other Training Aids

Participants' Manuals

The training delivered in the 1940s did not include a manual that would be given to each participant. Today I distribute a Participant's Manual for several reasons.

1. Many people like to take notes during a training session.
2. A completed manual gives each participant a record of the material to which s/he can refer after the training.
3. By incorporating some blanks in the material, the participants can gauge how much they know.
4. The manual can be used to hold blank forms so separate copying is not necessary.

The manual consists of an outline of the material delivered with occasional blanks in the sentences to check participants' knowledge. Reviewing the manuals as a group verifies that everyone has all the blanks filled in and thus that they have captured the basic concepts.

Videos

The durations of the demonstrations can vary, leaving some extra time available. Introducing videos also breaks up the pattern of training. In JIT I usually show two videos, both of which can be found on You Tube. "Problems In Supervision" is a training film made by the U.S. Department of Education in 1944. There are many learning points in this movie and you will probably see a different one almost every time you watch it. The main point is that the JBS is the heart of JIT and the reason the trainer in the movie does so well without one is that it is a movie and he is, in fact, following the script. If we want to deliver the same training every time, we must use the JBS, which is our "script."

The second movie is titled "The Invisible Gorilla," which is based on a psychology experiment. The point of this movie is that people often miss seeing something that we may think to be obvious. As a result, if you want someone to see something in your instruction, point it out to the learner and do not assume they see it. Looking is not the same as seeing.

Index

A

Auditing
   JMT, 130
   JRT, 149

B

BeLikeCoach organization, 253–258

C

Cascading, 73–79, 267–272
CI. *see* Continual improvement (CI) program
Continual improvement (CI) program, 104–106
Creativity function, as TWI, 249–251

D

Delivering instruction, JIT
   four-step method, 80–90
   job safety training, 90–92
   overview, 79–80
Developing JIT coaches, 184–186
Development of trainers
   JIT, 186–188
   JMT, 207
   JRT, 226–227
Documentation
   JIT, 95–96
   JMT, 130
   JRT, 149
   TWI, 175–176

F

The Fifth Discipline (Senge), 253
Foundational skills, as TWI, 251

G

Gap analysis, 155–157
Gino, Francesca, 243

H

Health and personal wellbeing, 251–252
HPT. *see* Human Performance Technology (HPT)
Human Performance Technology (HPT), 259

J

JBS. *see* Job breakdown sheet (JBS)
JMT. *see* Job methods training (JMT)
Job breakdown sheet (JBS), 42–44
   creating, 48–67, 287–289
   critiquing, 67–72
   finalizing, 72–73
   using, 289–292
Job instruction training (JIT). *see also* Training within Industry (TWI)
   advancing steps and key points, 44–47
   benefits of, 29–30
   beyond manufacturing, 92
   cascading, 73–79
   creating standard work, 31–32
   creation process, 92–96

auditing and updating, 93–95
documentation, 95–96
delivering instruction using
four-step method, 80–90
job safety training, 90–92
overview, 79–80
developing coaches, 184–186
developing JIT trainers, 186–188
format, 183–184
job breakdown sheet, 42–44
creating, 48–67
critiquing, 67–72
finalizing, 72–73
objectives of, 28–29
overview, 27
planning questionnaire, 297–298
preparation for using, 39–42
principles of, 32–39
standard work *versus* standardized work, 30–31
teaching, 295
terminology, 47–48
training
day 1, 189–195
day 2, 195–201
day 3, 201–202
day 4, 202–203
day 5, 203–204
delivering sessions, 188
use of, 204–206, 279–286
Job methods training (JMT)
auditing, 130
benefits of, 100–103
coaching, 207
continual improvement (CI) program, 104–106
developing trainers, 207
documentation, 130
format, 206–207
improvement methods, 224
misconceptions of, 103–104
objectives of, 99–100
principles of, 106–110
sessions, 207–223
day 1, 208–218
day 2, 218–223
day 3-5, 223
teaching, 295
usage of, 224
description, 114–122
proposal sheet, 122–126
viscosity check example, 126–129
Job relations training (JRT)
auditing and documentation, 149
benefits of, 134–135
changes to, 148–149
confidentiality, 228–232
developing trainers, 226–227
format, 225–226
improved relations, 232
objectives of, 133–134
principles of, 135–139
sessions, 227–228
teaching, 295
usage of, 139–148, 232
Job safety training, 90–92
Job selection and training room, 170–175
JRT. *see* Job relations training (JRT)

K

Kirkpatrick model, and TWI, 239–241

L

Learning Organization, 253, 273–277
Long-Term Program Development (LTPD), 258
LTPD. *see* Long-Term Program Development (LTPD)

M

Maslow's hierarchy, and TWI, 258–259

O

One-to-one instruction, 34–35

P

Pisano, Gary, 243
Process improvement *versus* problem-solving, 101
Proposal sheet, JMT, 122–126

R

Risk management, and TWI, 260

S

Scientific Method, 17
Self-determination theory, 273–277
Selling
   overview, 151–152
   standard actions
      initiator researches and TWI, 153–157
      managers accept small pilot program, 157–166
      overview, 152–153
Senge, Peter, 253
Skills Matrix. *see* Training Matrix
Standard work
   creating, 31–32
   versus standardized work, 30–31

T

Teamwork, 253
TGFS. *see* Toyota Gosei Fluid Systems (TGFS)
Toyota Gosei Fluid Systems (TGFS), 266
Training Matrix, 202, 263–266
Training within Industry (TWI). *see also specific types*
   from 1940–1945, 1–7
   and BeLikeCoach organization, 253–258
   beyond original intent, 249
   and cascading, 267–272
   confidence and resourcefulness, 20
   creativity function, 249–251
   embedding into culture, 241–243
   as foundational skills, 251
   general method, 21–23
   health and personal wellbeing, 251–252
   and Human Performance Technology (HPT), 259
   identify and solve problems, 294
   implementing and sustaining, 235–238
   and job breakdown sheet (JBS)
      creating, 287–289
      using, 289–292
   and Kirkpatrick model, 239–241
   lean foundation, 21
   and Learning Organization, 253, 273–277
   and Maslow's hierarchy, 258–259
   objective, 293
   and organization's operation, 24
   overview, 247–248
   and participants' manuals, 299
   and people's basic human needs, 24–25
   preparation
      documentation, assessments, and publications, 175–176
      finalize plan, 167
      general, 167–170
      job selection and training room, 170–175
   principles of
      bottom up *versus* top down, 17–18
      objectives, 18–20
      overview, 15–17
      Scientific Method, 17
   reintroduction of, 10–13
   and risk management, 260
   selling
      overview, 151–152
      standard actions, 152–166
   as skill based programs, 23
   successful use of, 244
   from 1945 to 2015, 7–10
   training, 176–182
      points common to J programs, 176–182
   Training Matrix, 263–266
   using three J programs together, 20
   and videos, 299–300
TWI. *see* Training within Industry (TWI)

V

Value Stream Mapping (VSM), 21
Viscosity check example, JMT, 126–129
VSM. *see* Value Stream Mapping (VSM)

W

Why Leaders Don't Learn from Success (Gino and Pisano), 243